A Philosophical Essay
on Probabilities

PIERRE SIMON,
MARQUIS DE LAPLACE

WITH AN INTRODUCTORY NOTE BY
E. T. BELL

DOVER PUBLICATIONS, INC.
New York

Bibliographical Note

This Dover edition, first published in 1995, is an unaltered and unabridged republication of the work originally published by John Wiley and Sons, New York, in 1902, and previously reprinted by Dover in 1952. The English translation, by Frederick William Truscott and Frederick Lincoln Emory, is from the sixth French edition of the work titled *Essai philosophique sur les probabilités,* published by Gauthier-Villars (Paris, n.d.) as part of the 15-volume series of Laplace's collected works. The original French edition was published in 1814 (Paris).

International Standard Book Number: 0-486-28875-7
Library of Congress Catalog Card Number: 52-12019

Manufactured in the United States of America
Dover Publications, Inc., 31 East 2nd Street, Mineola, N.Y. 11501

INTRODUCTORY NOTE

IN ADDITION to a vast mass of strictly technical writing in both pure and applied mathematics, Laplace (1749-1827) published two popular works addressed to the general intelligent and educated reader: *Exposition du système du monde* (1796), *Essai philosophique sur les probabilités,* only the second of which concerns us here. It formed the Introduction (153 pages) to the great, *Théorie analytique des probabilités* (645 pages), printed in Volume 7 of Laplace's collected works. The "avertissement" to the third edition (of the *Théorie,* 1820) states that the new edition differs from the preceding by, among other things, an introduction which had appeared as a separate publication "last year." This is the philosophical essay before us. In the "avertissement" to the second edition of the *Théorie,* Laplace expresses the hope that his labors will merit the attention of mathematicians and "excite" them to cultivate a branch of human knowledge as important and as curious as any. His hope has been realized; the mathematical theory of probability is indispensable today in both pure and applied science, and many of the tools still in use were invented by Laplace in his tremendous *Théorie,* the most impressive contribution as yet made by one man to this "curious and important branch of human knowledge."

The aim of the *Essai* apparently is to acquaint the reader with the fundamentals of probability and its applications without resorting to higher mathematics. Even elementary mathematics is kept to an irreducible minimum, as when, for example, it is necessary to refer to the binomial theorem for a positive integer exponent. But the simplicity is only apparent: mathematical operations, such as the multiplication of infinite series, remain mathematical although expressed in common language without any symbols whatever. For example, in the discussion of generating functions we are told to "multiply A by a function B of two variables, developed in a series arranged with respect to powers and products of powers of these variables, such, for example, as the first variable, plus the second, minus two," and so on for seven long lines of words. This is only a mild specimen of many similar attempts to make the essentially mathematical unmathematical. All such can be ignored without detriment to an understanding of the main purpose of the *Essai*—to acquaint the average reader with the concepts and uses of probability.

To place the *Essai* in its proper setting, we cannot do better than quote Laplace himself, and then see what his remarks were founded on. "This *Introduction*," he says, "is the development of a Lesson on Probabilities, which I gave in 1795 at the Normal Schools, where I was called as professor of Mathematics with Lagrange, by a decree of the National Convention. I was to present, without the aid of analysis [mathematics] the principles and general results of the theory of probabilities expounded in this work, in applying them to the most important questions of life, which are only in effect, for the most part, problems in probability." This explains the avoidance of formulas and technical mathematics already noted. Actually large tracts of the *Essai* are transpositions into common language of certain parts of the *Théorie analytique* developed mathematically and both more clearly and more succinctly than in the corresponding sections of the *Essai*. But even the reader who is acquainted with the formal mathematical development will find much to interest him in the informal presentation. To dispose here of the inevitable criticism that the theory of probability has grown and changed since Laplace's day—of course it has, but the origin of much of the growth and change is to be found in the theory as developed by him.

Neither the *Essai* nor the *Theorie* was composed at one sitting as it were. The final works were the outcome of numerous memoirs on special phases of the theory, dealing mostly with the relevant mathematics for which Laplace himself was largely responsible. If he did not actually originate all of the mathematics he used, he developed to a high pitch of power and applied it as nobody before him had. In the course of his work on probability he elaborated parts of a theory of equations in finite differences in one and in two independent variables. Among the applications of this analysis are the problems of duration of play, and the calculation of the chances that the number of counters in a random sample taken from a heap is odd or even. This investigation dates from 1774. From the year 1773 there is the basic investigation on inverse probabilities: to estimate the probabilities of the various causes that may have produced an observed event. This connects with the ideas of Thomas Boyes on inverse probability, published in 1763-4, and mentioned by Laplace. The subject long was controversial. At this early date Laplace was already applying the theory to questions in astronomy, such as the mean inclination of the orbits of comets. A memoir of 1781 contains much of mathematical interest, including the approximate evaluation of definite integrals. Passing to something more human, Laplace considers the probability that in any given year the number of boy-births will not exceed the number of girl-births. This suggests the corresponding and more difficult problem for a century.

We may note in passing the ingenious evaluation of $\int_0^\infty e^{-t^2}\,dt$, often given in texts on the calculus.

Another type of investigation deals with "the approximations of formulas which are functions of very large numbers."

This dates from 1782; the subject is still far from exhausted. With this memoir the first great period of Laplace's activity in the theory of probability closed. He did not significantly resume the subject for more than twenty-five years. But he was by no means idle; the greatest effort of his long and laborious career was being devoted to the composition of his masterpiece, the *Mécanique Céleste*. Then in 1809 he returned to the theory of probability with a memoir on the inclinations of the orbits of the planets and comets. The memoir contains further investigations in approximative calculation.

The foregoing sample of Laplace's contributions to the theory of probability before the publication of the definitive *Théorie Analytique*, is sufficient to show that the great work was the outcome of slow accretions through many years. In the final work much that had been obtained by laborious calculation was presented in simplified and improved form, and the many preliminaries were unified and drawn together. The massive effort was much greater than the sum of its parts, and is one of the outstanding examples in all mathematics of taking enough time to do properly a job worth doing at all. Finally, Laplace undertook to share his pleasure and his enthusiasm for the theory with those to whom the mathematics of the analytic theory are an impassable barrier. The result was the classic Essai that follows.

E. T. BELL

TABLE OF CONTENTS.

PART I.
A PHILOSOPHICAL ESSAY ON PROBABILITIES.

PART II.
APPLICATION OF THE CALCULUS OF PROBABILITIES.

A PHILOSOPHICAL ESSAY ON PROBABILITIES.

CHAPTER I.

INTRODUCTION.

THIS philosophical essay is the development of a lecture on probabilities which I delivered in 1795 to the normal schools whither I had been called, by a decree of the national convention, as professor of mathematics with Lagrange. I have recently published upon the same subject a work entitled *The Analytical Theory of Probabilities*. I present here without the aid of analysis the principles and general results of this theory, applying them to the most important questions of life, which are indeed for the most part only problems of probability. Strictly speaking it may even be said that nearly all our knowledge is problematical; and in the small number of things which we are able to know with certainty, even in the mathematical sciences themselves, the principal means for ascertaining truth —induction, and analogy—are based on probabilities;

so that the entire system of human knowledge is connected with the theory set forth in this essay. Doubtless it will be seen here with interest that in considering, even in the eternal principles of reason, justice, and humanity, only the favorable chances which are constantly attached to them, there is a great advantage in following these principles and serious inconvenience in departing from them: their chances, like those favorable to lotteries, always end by prevailing in the midst of the vacillations of hazard. I hope that the reflections given in this essay may merit the attention of philosophers and direct it to a subject so worthy of engaging their minds.

CHAPTER II.

CONCERNING PROBABILITY.

ALL events, even those which on account of their insignificance do not seem to follow the great laws of nature, are a result of it just as necessarily as the revolutions of the sun. In ignorance of the ties which unite such events to the entire system of the universe, they have been made to depend upon final causes or upon hazard, according as they occur and are repeated with regularity, or appear without regard to order; but these imaginary causes have gradually receded with the widening bounds of knowledge and disappear entirely before sound philosophy, which sees in them only the expression of our ignorance of the true causes.

Present events are connected with preceding ones by a tie based upon the evident principle that a thing cannot occur without a cause which produces it. This axiom, known by the name of *the principle of sufficient reason*, extends even to actions which are considered indifferent; the freest will is unable without a determinative motive to give them birth; if we assume two positions with exactly similar circumstances and find that the will is active in the one and inactive in the other, we say that its choice is an effect without a cause. It is then, says Leibnitz, the blind chance of the Epicureans. The contrary opinion is an illusion of the mind, which, losing sight of the evasive reasons of the choice of the will in indifferent things, believes that choice is determined of itself and without motives.

We ought then to regard the present state of the universe as the effect of its anterior state and as the cause of the one which is to follow. Given for one instant an intelligence which could comprehend all the forces by which nature is animated and the respective situation of the beings who compose it—an intelligence sufficiently vast to submit these data to analysis—it would embrace in the same formula the movements of the greatest bodies of the universe and those of the lightest atom; for it, nothing would be uncertain and the future, as the past, would be present to its eyes. The human mind offers, in the perfection which it has been able to give to astronomy, a feeble idea of this intelligence. Its discoveries in mechanics and geometry, added to that of universal gravity, have enabled it to comprehend in the same analytical expressions the past and future states of the system of the world. Applying the same method to some other objects of its knowledge, it has succeeded in referring to general laws observed phenomena and in foreseeing those which given circumstances ought to produce. All these efforts in the search for truth tend to lead it back continually to the vast intelligence which we have just mentioned, but from which it will always remain infinitely removed. This tendency, peculiar to the human race, is that which renders it superior to animals; and their progress

in this respect distinguishes nations and ages and con-
stitutes their true glory.

Let us recall that formerly, and at no remote epoch,
an unusual rain or an extreme drought, a comet having
in train a very long tail, the eclipses, the aurora
borealis, and in general all the unusual phenomena
were regarded as so many signs of celestial wrath.
Heaven was invoked in order to avert their baneful
influence. No one prayed to have the planets and the
sun arrested in their courses: observation had soon
made apparent the futility of such prayers. But as
these phenomena, occurring and disappearing at long
intervals, seemed to oppose the order of nature, it was
supposed that Heaven, irritated by the crimes of the
earth, had created them to announce its vengeance.
Thus the long tail of the comet of 1456 spread terror
through Europe, already thrown into consternation by
the rapid successes of the Turks, who had just over-
thrown the Lower Empire. This star after four revolutions has excited among us a very different interest.
The knowledge of the laws of the system of the world
acquired in the interval had dissipated the fears
begotten by the ignorance of the true relationship of
man to the universe; and Halley, having recognized
the identity of this comet with those of the years 1531,
1607, and 1682, announced its next return for the end
of the year 1758 or the beginning of the year 1759.
The learned world awaited with impatience this return
which was to confirm one of the greatest discoveries
that have been made in the sciences, and fulfil the
prediction of Seneca when he said, in speaking of the
revolutions of those stars which fall from an enormous

height: "The day will come when, by study pursued through several ages, the things now concealed will appear with evidence; and posterity will be astonished that truths so clear had escaped us." Clairaut then undertook to submit to analysis the perturbations which the comet had experienced by the action of the two great planets, Jupiter and Saturn; after immense calculations he fixed its next passage at the perihelion toward the beginning of April, 1759, which was actually verified by observation. The regularity which astronomy shows us in the movements of the comets doubtless exists also in all phenomena.

The curve described by a simple molecule of air or vapor is regulated in a manner just as certain as the planetary orbits; the only difference between them is that which comes from our ignorance.

Probability is relative, in part to this ignorance, in part to our knowledge. We know that of three or a greater number of events a single one ought to occur; but nothing induces us to believe that one of them will occur rather than the others. In this state of indecision it is impossible for us to announce their occurrence with certainty. It is, however, probable that one of these events, chosen at will, will not occur because we see several cases equally possible which exclude its occurrence, while only a single one favors it.

The theory of chance consists in reducing all the events of the same kind to a certain number of cases equally possible, that is to say, to such as we may be equally undecided about in regard to their existence, and in determining the number of cases favorable to the event whose probability is sought. The ratio of this number to that of all the cases possible is the measure of this probability, which is thus simply a fraction whose numerator is the number of favorable cases and whose denominator is the number of all the cases possible.

The preceding notion of probability supposes that, in increasing in the same ratio the number of favorable cases and that of all the cases possible, the probability remains the same. In order to convince ourselves let us take two urns, A and B, the first containing four white and two black balls, and the second containing only two white balls and one black one. We may imagine the two black balls of the first urn attached by a thread which breaks at the moment when one of them is seized in order to be drawn out, and the four white balls thus forming two similar systems. All the chances which will favor the seizure of one of the balls of the black system will lead to a black ball. If we conceive now that the threads which unite the balls do not break at all, it is clear that the number of possible chances will not change any more than that of the chances favorable to the extraction of the black balls; but two balls will be drawn from the urn at the same time; the probability of drawing a black ball from the urn A will then be the same as at first. But then we have obviously the case of urn B with the single difference that the three balls of this last urn would be replaced by three systems of two balls invariably connected.

When all the cases are favorable to an event the probability changes to certainty and its expression becomes equal to unity. Upon this condition, certainty and probability are comparable, although there may be an essential difference between the two states of the mind when a truth is rigorously demonstrated to it, or when it still perceives a small source of error.

In things which are only probable the difference of the data, which each man has in regard to them, is one of the principal causes of the diversity of opinions which prevail in regard to the same objects. Let us suppose, for example, that we have three urns, A, B, C, one of which contains only black balls while the two others contain only white balls; a ball is to be drawn from the urn C and the probability is demanded that this ball will be black. If we do not know which of the three urns contains black balls only, so that there is no reason to believe that it is C rather than B or A, these three hypotheses will appear equally possible, and since a black ball can be drawn only in the first hypothesis, the probability of drawing it is equal to one third. If it is known that the urn A contains white balls only, the indecision then extends only to the urns B and C, and the probability that the ball drawn from the urn C will be black is one half. Finally this probability changes to certainty if we are assured that the urns A and B contain white balls only.

It is thus that an incident related to a numerous assembly finds various degrees of credence, according to the extent of knowledge of the auditors. If the man who reports it is fully convinced of it and if, by his position and character, he inspires great confidence, his statement, however extraordinary it may be, will have for the auditors who lack information the same degree of probability as an ordinary statement made by the same man, and they will have entire faith in it. But if some one of them knows that the same incident is rejected by other equally trustworthy men, he will be in doubt and the incident will be discredited by the enlightened auditors, who will reject it whether it be in regard to facts well averred or the immutable laws of nature.

It is to the influence of the opinion of those whom the multitude judges best informed and to whom it has been accustomed to give its confidence in regard to the most important matters of life that the propagation of those errors is due which in times of ignorance have covered the face of the earth. Magic and astrology offer us two great examples. These errors inculcated in infancy, adopted without examination, and having for a basis only universal credence, have maintained themselves during a very long time; but at last the progress of science has destroyed them in the minds of enlightened men, whose opinion consequently has caused them to disappear even among the common people, through the power of imitation and habit which had so generally spread them abroad. This power, the richest resource of the moral world, establishes and conserves in a whole nation ideas entirely contrary to those which it upholds elsewhere with the same authority. What indulgence ought we not then to have for opinions different from ours, when this difference often depends only upon the various points of view where circumstances have placed us! Let us enlighten those whom we judge insufficiently instructed; but first let us examine critically our own opinions and weigh with impartiality their respective probabilities.

The difference of opinions depends, however, upon the manner in which the influence of known data is determined. The theory of probabilities holds to considerations so delicate that it is not surprising that with the same data two persons arrive at different results, especially in very complicated questions. Let us examine now the general principles of this theory.

CHAPTER III.

THE GENERAL PRINCIPLES OF THE CALCULUS OF PROBABILITIES.

First Principle.—The first of these principles is the definition itself of probability, which, as has been seen, is the ratio of the number of favorable cases to that of all the cases possible.

Second Principle.—But that supposes the various cases equally possible. If they are not so, we will determine first their respective possibilities, whose exact appreciation is one of the most delicate points of the theory of chance. Then the probability will be the sum of the possibilities of each favorable case. Let us illustrate this principle by an example.

Let us suppose that we throw into the air a large and very thin coin whose two large opposite faces, which we will call heads and tails, are perfectly similar. Let us find the probability of throwing heads at least one time in two throws. It is clear that four equally possible cases may arise, namely, heads at the first and at the second throw; heads at the first throw and tails at the second; tails at the first throw and heads at the second; finally, tails at both throws. The first three cases are favorable to the event whose probability is sought; consequently this probability is equal to $\frac{3}{4}$; so that it is a bet of three to one that heads will be thrown at least once in two throws.

We can count at this game only three different cases, namely, heads at the first throw, which dispenses with throwing a second time; tails at the first throw and heads at the second; finally, tails at the first and at the second throw. This would reduce the probability to $\frac{2}{3}$ if we should consider with d'Alembert these three cases as equally possible. But it is apparent that the probability of throwing heads at the first throw is $\frac{1}{2}$, while that of the two other cases is $\frac{1}{4}$, the first case being a simple event which corresponds to two events combined: heads at the first and at the second throw, and heads at the first throw, tails at the second. If we then, conforming to the second principle, add the possibility $\frac{1}{2}$ of heads at the first throw to the possibility $\frac{1}{4}$ of tails at the first throw and heads at the second, we shall have $\frac{3}{4}$ for the probability sought, which agrees with what is found in the supposition when we play the two throws. This supposition does not change at all the chance of that one who bets on this event; it simply serves to reduce the various cases to the cases equally possible.

Third Principle.—One of the most important points of the theory of probabilities and that which lends the most to illusions is the manner in which these probabilities increase or diminish by their mutual combination. If the events are independent of one another, the probability of their combined existence is the product of their respective probabilities. Thus the probability of throwing one ace with a single die is $\frac{1}{6}$; that of throwing two aces in throwing two dice at the same time is $\frac{1}{36}$. Each face of the one being able to combine with the six faces of the other, there are in fact thirty-six equally possible cases, among which one single case gives two aces. Generally the probability that a simple event in the same circumstances will occur consecutively a given number of times is equal to the probability of this simple event raised to the power indicated by this number. Having thus the successive powers of a fraction less than unity diminishing without ceasing, an event which depends upon a series of very great probabilities may become extremely improbable. Suppose then an incident be transmitted to us by twenty witnesses in such manner that the first has transmitted it to the second, the second to the third, and so on. Suppose again that the probability of each testimony be equal to the fraction $\frac{9}{10}$; that of the incident resulting from the testimonies will be less than $\frac{1}{8}$. We cannot better compare this diminution of the probability than with the extinction of the light of objects by the interposition of several pieces of glass. A relatively small number of pieces suffices to take away the view of an object that a single piece allows us to perceive in a distinct manner. The historians do not appear to have paid sufficient attention to this degradation of the probability of events when seen across a great number of successive generations; many historical events reputed as certain would be at least doubtful if they were submitted to this test.

In the purely mathematical sciences the most distant consequences participate in the certainty of the principle from which they are derived.   In the applications of analysis to physics the results have all the certainty of facts or experiences.   But in the moral sciences, where each inference is deduced from that which precedes it only in a probable manner, however probable these deductions may be, the chance of error increases with their number and ultimately surpasses the chance of truth in the consequences very remote from the principle.

Fourth Principle.—When two events depend upon each other, the probability of the compound event is the product of the probability of the first event and the probability that, this event having occurred, the second will occur.   Thus in the preceding case of the three urns A, B, C, of which two contain only white balls and one contains only black balls, the probability of drawing a white ball from the urn C is $\frac{2}{3}$, since of the three urns only two contain balls of that color.   But when a white ball has been drawn from the urn C, the indecision relative to that one of the urns which contain only black balls extends only to the urns A and B; the probability of drawing a white ball from the urn B is $\frac{1}{2}$; the product of $\frac{2}{3}$ by $\frac{1}{2}$, or $\frac{1}{3}$, is then the probability of drawing two white balls at one time from the urns B and C.

We see by this example the influence of past events upon the probability of future events.   For the probability of drawing a white ball from the urn B, which primarily is $\frac{2}{3}$, becomes $\frac{1}{2}$ when a white ball has been drawn from the urn C; it would change to certainty if a black ball had been drawn from the same urn.   We will determine this influence by means of the following principle, which is a corollary of the preceding one.

Fifth Principle.—If we calculate *à priori* the probability of the occurred event and the probability of an event composed of that one and a second one which is expected, the second probability divided by the first will be the probability of the event expected, drawn from the observed event.

Here is presented the question raised by some philosophers touching the influence of the past upon the probability of the future. Let us suppose at the play of heads and tails that heads has occurred oftener than tails. By this alone we shall be led to believe that in the constitution of the coin there is a secret cause which favors it. Thus in the conduct of life constant happiness is a proof of competency which should induce us to employ preferably happy persons. But if by the unreliability of circumstances we are constantly brought back to a state of absolute indecision, if, for example, we change the coin at each throw at the play of heads and tails, the past can shed no light upon the future and it would be absurd to take account of it.

Sixth Principle.—Each of the causes to which an observed event may be attributed is indicated with just as much likelihood as there is probability that the event will take place, supposing the event to be constant. The probability of the existence of any one of these causes is then a fraction whose numerator is the probability of the event resulting from this cause and whose denominator is the sum of the similar probabilities relative to all the causes; if these various causes, considered *à priori*, are unequally probable, it is necessary, in place of the probability of the event resulting from each cause, to employ the product of this probability by the possibility of the cause itself.   This is the fundamental principle of this branch of the analysis of chances which consists in passing from events to causes.

This principle gives the reason why we attribute regular events to a particular cause.   Some philosophers have thought that these events are less possible than others and that at the play of heads and tails, for example, the combination in which heads occurs twenty successive times is less easy in its nature than those where heads and tails are mixed in an irregular manner. But this opinion supposes that past events have an influence on the possibility of future events, which is not at all admissible.   The regular combinations occur more rarely only because they are less numerous.   If we seek a cause wherever we perceive symmetry, it is not that we regard a symmetrical event as less possible than the others, but, since this event ought to be the effect of a regular cause or that of chance, the first of these suppositions is more probable than the second. On a table we see letters arranged in this order, *C o n s t a n t i n o p l e*, and we judge that this arrangement is not the result of chance, not because it is less possible than the others, for if this word were not employed in any language we should not suspect it came from any particular cause, but this word being in use among us, it is incomparably more probable that some person has thus arranged the aforesaid letters than that this arrangement is due to chance.

This is the place to define the word *extraordinary*. We arrange in our thought all possible events in various classes; and we regard as *extraordinary* those classes which include a very small number. Thus at the play of heads and tails the occurrence of heads a hundred successive times appears to us extraordinary because of the almost infinite number of combinations which may occur in a hundred throws; and if we divide the combinations into regular series containing an order easy to comprehend, and into irregular series, the latter are incomparably more numerous. The drawing of a white ball from an urn which among a million balls contains only one of this color, the others being black, would appear to us likewise extraordinary, because we form only two classes of events relative to the two colors. But the drawing of the number 475813, for example, from an urn that contains a million numbers seems to us an ordinary event; because, comparing individually the numbers with one another without dividing them into classes, we have no reason to believe that one of them will appear sooner than the others.

From what precedes, we ought generally to conclude that the more extraordinary the event, the greater the need of its being supported by strong proofs. For those who attest it, being able to deceive or to have been deceived, these two causes are as much more probable as the reality of the event is less. We shall see this particularly when we come to speak of the probability of testimony.

Seventh Principle.—The probability of a future event is the sum of the products of the probability of each cause, drawn from the event observed, by the probability that, this cause existing, the future event will occur. The following example will illustrate this principle.

Let us imagine an urn which contains only two balls, each of which may be either white or black. One of these balls is drawn and is put back into the urn before proceeding to a new draw. Suppose that in the first two draws white balls have been drawn; the probability of again drawing a white ball at the third draw is required.

Only two hypotheses can be made here: either one of the balls is white and the other black, or both are white. In the first hypothesis the probability of the event observed is $\frac{1}{4}$; it is unity or certainty in the second. Thus in regarding these hypotheses as so many causes, we shall have for the sixth principle $\frac{1}{5}$ and $\frac{4}{5}$ for their respective probabilities. But if the first hypothesis occurs, the probability of drawing a white ball at the third draw is $\frac{1}{2}$; it is equal to certainty in the second hypothesis; multiplying then the last probabilities by those of the corresponding hypotheses, the sum of the products, or $\frac{9}{10}$, will be the probability of drawing a white ball at the third draw.

When the probability of a single event is unknown we may suppose it equal to any value from zero to unity. The probability of each of these hypotheses, drawn from the event observed, is, by the sixth principle, a fraction whose numerator is the probability of the event in this hypothesis and whose denominator is the sum of the similar probabilities relative to all the hypotheses. Thus the probability that the possibility of the event is comprised within given limits is the sum of the fractions comprised within these limits. Now if we multiply each fraction by the probability of the future event, determined in the corresponding hypothesis, the sum of the products relative to all the hypotheses will be, by the seventh principle, the probability of the future event drawn from the event observed. Thus we find that an event having occurred successively any number of times, the probability that it will happen again the next time is equal to this number increased by unity divided by the same number, increased by two units. Placing the most ancient epoch of history at five thousand years ago, or at 182623 days, and the sun having risen constantly in the interval at each revolution of twenty-four hours, it is a bet of 1826214 to one that it will rise again to-morrow. But this number is incomparably greater for him who, recognizing in the totality of phenomena the principal regulator of days and seasons, sees that nothing at the present moment can arrest the course of it.

Buffon in his *Political Arithmetic* calculates differently the preceding probability. He supposes that it differs from unity only by a fraction whose numerator is unity and whose denominator is the number 2 raised to a power equal to the number of days which have elapsed since the epoch. But the true manner of relating past events with the probability of causes and of future events was unknown to this illustrious writer.

CHAPTER IV.

CONCERNING HOPE.

THE probability of events serves to determine the hope or the fear of persons interested in their existence. The word *hope* has various acceptations; it expresses generally the advantage of that one who expects a certain benefit in suppositions which are only probable. This advantage in the theory of chance is a product of the sum hoped for by the probability of obtaining it; it is the partial sum which ought to result when we do not wish to run the risks of the event in supposing that the division is made proportional to the probabilities. This division is the only equitable one when all strange circumstances are eliminated; because an equal degree of probability gives an equal right to the sum hoped for. We will call this advantage *mathematical hope*.

Eighth Principle.—When the advantage depends on several events it is obtained by taking the sum of the products of the probability of each event by the benefit attached to its occurrence.

Let us apply this principle to some examples. Let

us suppose that at the play of heads and tails Paul receives two francs if he throws heads at the first throw and five francs if he throws it only at the second. Multiplying two francs by the probability $\frac{1}{2}$ of the first case, and five francs by the probability $\frac{1}{4}$ of the second case, the sum of the products, or two and a quarter francs, will be Paul's advantage. It is the sum which he ought to give in advance to that one who has given him this advantage; for, in order to maintain the equality of the play, the throw ought to be equal to the advantage which it procures.

If Paul receives two francs by throwing heads at the first and five francs by throwing it at the second throw, whether he has thrown it or not at the first, the probability of throwing heads at the second throw being $\frac{1}{2}$, multiplying two francs and five francs by $\frac{1}{2}$ the sum of these products will give three and one half francs for Paul's advantage and consequently for his stake at the game.

Ninth Principle.—In a series of probable events of which the ones produce a benefit and the others a loss, we shall have the advantage which results from it by making a sum of the products of the probability of each favorable event by the benefit which it procures, and subtracting from this sum that of the products of the probability of each unfavorable event by the loss which is attached to it. If the second sum is greater than the first, the benefit becomes a loss and hope is changed to fear.

Consequently we ought always in the conduct of life to make the product of the benefit hoped for, by its probability, at least equal to the similar product relative to the loss. But it is necessary, in order to attain this, to appreciate exactly the advantages, the losses, and their respective probabilities. For this a great accuracy of mind, a delicate judgment, and a great experience in affairs is necessary; it is necessary to know how to guard one's self against prejudices, illusions of fear or hope, and erroneous ideas, ideas of fortune and happiness, with which the majority of people feed their self-love.

The application of the preceding principles to the following question has greatly exercised the geometricians. Paul plays at heads and tails with the condition of receiving two francs if he throws heads at the first throw, four francs if he throws it only at the second throw, eight francs if he throws it only at the third, and so on. His stake at the play ought to be, according to the eighth principle, equal to the number of throws, so that if the game continues to infinity the stake ought to be infinite. However, no reasonable man would wish to risk at this game even a small sum, for example five francs. Whence comes this difference between the result of calculation and the indication of common sense ? We soon recognize that it amounts to this: that the moral advantage which a benefit procures for us is not proportional to this benefit and that it depends upon a thousand circumstances, often very difficult to define, but of which the most general and most important is that of fortune.

Indeed it is apparent that one franc has much greater value for him who possesses only a hundred than for a millionaire. We ought then to distinguish in the hoped-for benefit its absolute from its relative value.

But the latter is regulated by the motives which make it desirable, whereas the first is independent of them. The general principle for appreciating this relative value cannot be given, but here is one proposed by Daniel Bernoulli which will serve in many cases.

Tenth Principle.—The relative value of an infinitely small sum is equal to its absolute value divided by the total benefit of the person interested. This supposes that every one has a certain benefit whose value can never be estimated as zero. Indeed even that one who possesses nothing always gives to the product of his labor and to his hopes a value at least equal to that which is absolutely necessary to sustain him.

If we apply analysis to the principle just propounded, we obtain the following rule: Let us designate by unity the part of the fortune of an individual, independent of his expectations. If we determine the different values that this fortune may have by virtue of these expectations and their probabilities, the product of these values raised respectively to the powers indicated by their probabilities will be the physical fortune which would procure for the individual the same moral advantage which he receives from the part of his fortune taken as unity and from his expectations; by subtracting unity from the product, the difference will be the increase of the physical fortune due to expectations: we will call this increase *moral hope.* It is easy to see that it coincides with mathematical hope when the fortune taken as unity becomes infinite in reference to the variations which it receives from the expectations. But when these variations are an appreciable part of this unity the two hopes may differ very materially among themselves.

This rule conduces to results conformable to the indications of common sense which can by this means be appreciated with some exactitude. Thus in the preceding question it is found that if the fortune of Paul is two hundred francs, he ought not reasonably to stake more than nine francs. The same rule leads us again to distribute the danger over several parts of a benefit expected rather than to expose the entire benefit to this danger. It results similarly that at the fairest game the loss is always greater than the gain. Let us suppose, for example, that a player having a fortune of one hundred francs risks fifty at the play of heads and tails; his fortune after his stake at the play will be reduced to eighty-seven francs, that is to say, this last sum would procure for the player the same moral advantage as the state of his fortune after the stake. The play is then disadvantageous even in the case where the stake is equal to the product of the sum hoped for, by its probability. We can judge by this of the immorality of games in which the sum hoped for is below this product. They subsist only by false reasonings and by the cupidity which they excite and which, leading the people to sacrifice their necessaries to chimerical hopes whose improbability they are not in condition to appreciate, are the source of an infinity of evils.

The disadvantage of games of chance, the advantage of not exposing to the same danger the whole benefit that is expected, and all the similar results indicated by common sense, subsist, whatever may be the function of the physical fortune which for each individual expresses his moral fortune. It is enough that the proportion of the increase of this function to the increase of the physical fortune diminishes in the measure that the latter increases.

CHAPTER V.

CONCERNING THE ANALYTICAL METHODS OF THE CALCULUS OF PROBABILITIES.

THE application of the principle which we have just expounded to the various questions of probability requires methods whose investigation has given birth to several methods of analysis and especially to the theory of combinations and to the calculus of finite differences.

If we form the product of the binomials, unity plus the first letter, unity plus the second letter, unity plus the third letter, and so on up to n letters, and subtract unity from this developed product, the result will be the sum of the combination of all these letters taken one by one, two by two, three by three, etc., each combination having unity for a coefficient. In order to have the number of combinations of these n letters taken s by s times, we shall observe that if we suppose these letters equal among themselves, the preceding product will become the nth power of the binomial one plus the first letter; thus the number of combinations of n letters taken s by s times will be the coefficient of the sth power of the first letter in the development in this binomial; and this number is obtained by means of the known binomial formula.

Attention must be paid to the respective situations of the letters in each combination, observing that if a second letter is joined to the first it may be placed in the first or second position which gives two combinations. If we join to these combinations a third letter, we can give it in each combination the first, the second, and the third rank which forms three combinations relative to each of the two others, in all six combinations. From this it is easy to conclude that the number of arrangements of which *s* letters are susceptible is the product of the numbers from unity to *s*. In order to pay regard to the respective positions of the letters it is necessary then to multiply by this product the number of combinations of *n* letters *s* by *s* times, which is tantamount to taking away the denominator of the coefficient of the binomial which expresses this number.

Let us imagine a lottery composed of *n* numbers, of which *r* are drawn at each draw. The probability is demanded of the drawing of *s* given numbers in one draw. To arrive at this let us form a fraction whose denominator will be the number of all the cases possible or of the combinations of *n* letters taken *r* by *r* times, and whose numerator will be the number of all the combinations which contain the given *s* numbers. This last number is evidently that of the combinations of the other numbers taken *n* less *s* by *n* less *s* times. This fraction will be the required probability, and we shall easily find that it can be reduced to a fraction whose numerator is the number of combinations of *r* numbers taken s by s times, and whose denominator is the number of combinations of n numbers taken similarly s by s times.   Thus in the lottery of France, formed as is known of 90 numbers of which five are drawn at each draw, the probability of drawing a given combination is $\frac{5}{90}$, or $\frac{1}{18}$; the lottery ought then for the equality of the play to give eighteen times the stake. The total number of combinations two by two of the 90 numbers is 4005, and that of the combinations two by two of 5 numbers is 10.   The probability of the drawing of a given pair is then $\frac{1}{4005}$, and the lottery ought to give four hundred and a half times the stake; it ought to give 11748 times for a given tray, 511038 times for a quaternary, and 43949268 times for a quint. The lottery is far from giving the player these advantages.

Suppose in an urn a white balls, b black balls, and after having drawn a ball it is put back into the urn; the probability is asked that in n number of draws m white balls and $n - m$ black balls will be drawn.   It is clear that the number of cases that may occur at each drawing is $a + b$.   Each case of the second drawing being able to combine with all the cases of the first, the number of possible cases in two drawings is the square of the binomial $a + b$.   In the development of this square, the square of a expresses the number of cases in which a white ball is twice drawn, the double product of a by b expresses the number of cases in which a white ball and a black ball are drawn.   Finally, the square of b expresses the number of cases in which two black balls are drawn.   Continuing thus, we see generally that the nth power of the binomial $a + b$

expresses the number of all the cases possible in n
draws; and that in the development of this power the
term multiplied by the mth power of a expresses the
number of cases in which m white balls and $n - m$
black balls may be drawn.   Dividing then this term
by the entire power of the binomial, we shall have the
probability of drawing m white balls and $n - m$ black
balls.   The ratio of the numbers a and $a + b$ being
the probability of drawing one white ball at one draw;
and the ratio of the numbers b and $a + b$ being the
probability of drawing one black ball; if we call these
probabilities p and q, the probability of drawing m white
balls in n draws will be the term multiplied by the mth
power of p in the development of the nth power of the
binomial $p + q$; we may see that the sum $p + q$ is
unity.   This remarkable property of the binomial is
very useful in the theory of probabilities.   But the
most general and direct method of resolving questions
of probability consists in making them depend upon
equations of differences.   Comparing the successive
conditions of the function which expresses the probability when we increase the variables by their respective differences, the proposed question often furnishes a
very simple proportion between the conditions.   This
proportion is what is called *equation of ordinary or
partial differentials; ordinary* when there is only one
variable, *partial* when there are several.   Let us consider some examples of this.

Three players of supposed equal ability play together
on the following conditions: that one of the first two
players who beats his adversary plays the third, and if
he beats him the game is finished.   If he is beaten, the

victor plays against the second until one of the players has defeated consecutively the two others, which ends the game. The probability is demanded that the game will be finished in a certain number n of plays. Let us find the probability that it will end precisely at the nth play. For that the player who wins ought to enter the game at the play $n - 1$ and win it thus at the following play. But if in place of winning the play $n - 1$ he should be beaten by his adversary who had just beaten the other player, the game would end at this play. Thus the probability that one of the players will enter the game at the play $n - 1$ and will win it is equal to the probability that the game will end precisely with this play; and as this player ought to win the following play in order that the game may be finished at the nth play, the probability of this last case will be only one half of the preceding one. This probability is evidently a function of the number n; this function is then equal to the half of the same function when n is diminished by unity. This equality forms one of those equations called *ordinary finite differential equations*.

We may easily determine by its use the probability that the game will end precisely at a certain play. It is evident that the play cannot end sooner than at the second play; and for this it is necessary that that one of the first two players who has beaten his adversary should beat at the second play the third player; the probability that the game will end at this play is $\frac{1}{2}$. Hence by virtue of the preceding equation we conclude that the successive probabilities of the end of the game are $\frac{1}{4}$ for the third play, $\frac{1}{8}$ for the fourth play, and so

on; and in general $\frac{1}{2}$ raised to the power $n - 1$ for the
nth play. The sum of all these powers of $\frac{1}{2}$ is unity
less the last of these powers; it is the probability that
the game will end at the latest in n plays.

Let us consider again the first problem more difficult
which may be solved by probabilities and which Pascal
proposed to Fermat to solve. Two players, A and B,
of equal skill play together on the conditions that the
one who first shall beat the other a given number of
times shall win the game and shall take the sum of the
stakes at the game; after some throws the players
agree to quit without having finished the game: we ask
in what manner the sum ought to be divided between
them. It is evident that the parts ought to be proportional to the respective probabilities of winning the
game. The question is reduced then to the determina-
tion of these probabilities. They depend evidently
upon the number of points which each player lacks of
having attained the given number. Hence the probability of A is a function of the two numbers which we
will call *indices.* If the two players should agree to
play one throw more (an agreement which does not
change their condition, provided that after this new
throw the division is always made proportionally to the
new probabilities of winning the game), then either A
would win this throw and in that case the number of
points which he lacks would be diminished by unity,
or the player B would win it and in that case the
number of points lacking to this last player would be
less by unity. But the probability of each of these
cases is $\frac{1}{2}$; the function sought is then equal to one half
of this function in which we diminish by unity the first

index plus the half of the same function in which the second variable is diminished by unity. This equality is one of those equations called *equations of partial differentials.*

We are able to determine by its use the probabilities of A by dividing the smallest numbers, and by observing that the probability or the function which expresses it is equal to unity when the player A does not lack a single point, or when the first index is zero, and that this function becomes zero with the second index. Supposing thus that the player A lacks only one point, we find that his probability is $\frac{1}{2}$, $\frac{3}{4}$, $\frac{7}{8}$, etc., according as B lacks one point, two, three, etc. Generally it is then unity less the power of $\frac{1}{2}$, equal to the number of points which B lacks. We will suppose then that the player A lacks two points and his probability will be found equal to $\frac{1}{4}$, $\frac{1}{2}$, $\frac{11}{16}$, etc., according as B lacks one point, two points, three points, etc. We will suppose again that the player A lacks three points, and so on.

This manner of obtaining the successive values of a quantity by means of its equation of differences is long and laborious. The geometricians have sought methods to obtain the general function of indices that satisfies this equation, so that for any particular case we need only to substitute in this function the corresponding values of the indices. Let us consider this subject in a general way. For this purpose let us conceive a series of terms arranged along a horizontal line so that each of them is derived from the preceding one according to a given law. Let us suppose this law expressed by an equation among several consecutive terms and their index, or the number which indicates the rank that they occupy in the series. This equation I call the *equation of finite differences by a single index.* The order or the degree of this equation is the difference of rank of its two extreme terms. We are able by its use to determine successively the terms of the series and to continue it indefinitely; but for that it is necessary to know a number of terms of the series equal to the degree of the equation. These terms are the arbitrary constants of the expression of the general term of the series or of the integral of the equation of differences.

Let us imagine now below the terms of the preceding series a second series of terms arranged horizontally; let us imagine again below the terms of the second series a third horizontal series, and so on to infinity; and let us suppose the terms of all these series connected by a general equation among several consecutive terms, taken as much in the horizontal as in the vertical sense, and the numbers which indicate their rank in the two senses. This equation is called the *equation of partial finite differences by two indices.*

Let us imagine in the same way below the plan of the preceding series a second plan of similar series, whose terms should be placed respectively below those of the first plan; let us imagine again below this second plan a third plan of similar series, and so on to infinity; let us suppose all the terms of these series connected by an equation among several consecutive terms taken in the sense of length, width, and depth, and the three numbers which indicate their rank in these three senses. This equation I call the *equation of partial finite differences by three indices.*

Finally, considering the matter in an abstract way

and independently of the dimensions of space, let us imagine generally a system of magnitudes, which should be functions of a certain number of indices, and let us suppose among these magnitudes, their relative differences to these indices and the indices themselves, as many equations as there are magnitudes; these equations will be partial finite differences by a certain number of indices.

We are able by their use to determine successively these magnitudes. But in the same manner as the equation by a single index requires for it that we known a certain number of terms of the series, so the equation by two indices requires that we know one or several lines of series whose general terms should be expressed each by an arbitrary function of one of the indices. Similarly the equation by three indices requires that we know one or several plans of series, the general terms of which should be expressed each by an arbitrary function of two indices, and so on. In all these cases we shall be able by successive eliminations to determine a certain term of the series. But all the equations among which we eliminate being comprised in the same system of equations, all the expressions of the successive terms which we obtain by these eliminations ought to be comprised in one general expression, a function of the indices which determine the rank of the term. This expression is the integral of the proposed equation of differences, and the search for it is the object of integral calculus.

Taylor is the first who in his work entitled *Metodus incrementorum* has considered linear equations of finite differences. He gives the manner of integrating those of the first order with a coefficient and a last term, functions of the index. In truth the relations of the terms of the arithmetical and geometrical progressions which have always been taken into consideration are the simplest cases of linear equations of differences; but they had not been considered from this point of view. It was one of those which, attaching themselves to general theories, lead to these theories and are consequently veritable discoveries.

About the same time Moivre was considering under the name of recurring series the equations of finite differences of a certain order having a constant coefficient. He succeeded in integrating them in a very ingenious manner. As it is always interesting to follow the progress of inventors, I shall expound the method of Moivre by applying it to a recurring series whose relation among three consecutive terms is given. First he considers the relation among the consecutive terms of a geometrical progression or the equation of two terms which expresses it. Referring it to terms less than unity, he multiplies it in this state by a constant factor and subtracts the product from the first equation. Thus he obtains an equation among three consecutive terms of the geometrical progression. Moivre considers next a second progression whose ratio of terms is the same factor which he has just used. He diminishes similarly by unity the index of the terms of the equation of this new progression. In this condition he multiplies it by the ratio of the terms of the first progression, and he subtracts the product from the equation of the second progression, which gives him among three consecutive terms of this progression a relation entirely similar to that which he has found for the first progression. Then he observes that if one adds term by term the two progressions, the same ratio exists among any three of these consecutive terms. He compares the coefficients of this ratio to those of the relation of the terms of the proposed recurrent series, and he finds for determining the ratios of the two geometrical progressions an equation of the second degree, whose roots are these ratios. Thus Moivre decomposes the recurrent series into two geometrical progressions, each multiplied by an arbitrary constant which he determines by means of the first two terms of the recurrent series. This ingenious process is in fact the one that d'Alembert has since employed for the integration of linear equations of infinitely small differences with constant coefficients, and Lagrange has transformed into similar equations of finite differences.

Finally, I have considered the linear equations of partial finite differences, first under the name of *recurro-recurrent* series and afterwards under their own name. The most general and simplest manner of integrating all these equations appears to me that which I have based upon the consideration of discriminant functions, the idea of which is here given.

If we conceive a function V of a variable t developed according to the powers of this variable, the coefficient of any one of these powers will be a function of the exponent or index of this power, which index I shall call x. V is what I call the discriminant function of this coefficient or of the function of the index.

Now if we multiply the series of the development of V by a function of the same variable, such, for example,

as unity plus two times this variable, the product will
be a new discriminant function in which the coefficient
of the power x of the variable t will be equal to the
coefficient of the same power in V plus twice the
coefficient of the power less unity. Thus the function
of the index x in the product will be equal to the function of the index x in V plus twice the same function
in which the index is diminished by unity. This function of the index x is thus a derivative of the function
of the same index in the development of V, a function
which I shall call the *primitive function* of the index.
Let us designate the derivative function by the letter δ
placed before the primitive function. The derivation
indicated by this letter will depend upon the multiplier
of V, which we will call T and which we will suppose
developed like V by the ratio to the powers of the
variable t. If we multiply anew by T the product of
V by T, which is equivalent to multiplying V by T^2,
we shall form a third discriminant function, in which
the coefficient of the xth power of t will be a derivative
similar to the corresponding coefficient of the preceding
product; it may be expressed by the same character δ
placed before the preceding derivative, and then this
character will be written twice before the primitive
function of x. But in place of writing it thus twice we
give it 2 for an exponent.

Continuing thus, we see generally that if we multiply
V by the nth power of T, we shall have the coefficient
of the xth power of t in the product of V by the nth
power of T by placing before the primitive function the
character δ with n for an exponent.

Let us suppose, for example, that T be unity divided

by t; then in the product of V by T the coefficient of the xth power of t will be the coefficient of the power greater by unity in V; this coefficient in the product of V by the nth power of T will then be the primitive function in which x is augmented by n units.

Let us consider now a new function Z of t, developed like V and T according to the powers of t; let us designate by the character $\varDelta$ placed before the primitive function the coefficient of the xth power of t in the product of V by Z; this coefficient in the product of V by the nth power of Z will be expressed by the character $\varDelta$ affected by the exponent n and placed before the primitive function of x.

If, for example, Z is equal to unity divided by t less one, the coefficient of the xth power of t in the product of V by Z will be the coefficient of the $x + 1$ power of t in V less the coefficient of the xth power. It will be then the finite difference of the primitive function of the index x. Then the character $\varDelta$ indicates a finite difference of the primitive function in the case where the index varies by unity; and the nth power of this character placed before the primitive function will indicate the finite nth difference of this function. If we suppose that T be unity divided by t, we shall have T equal to the binomial $Z + 1$. The product of V by the nth power of T will then be equal to the product of V by the nth power of the binomial $Z + 1$. Developing this power in the ratio of the powers of Z, the product of V by the various terms of this development will be the discriminant functions of these same terms in which we substitute in place of the powers of Z the corresponding finite differences of the primitive function of the index.

Now the product of V by the nth power of T is the primitive function in which the index x is augmented by n units; repassing from the discriminant functions to their coefficients, we shall have this primitive function thus augmented equal to the development of the nth power of the binomial $Z + 1$, provided that in this development we substitute in place of the powers of Z the corresponding differences of the primitive function and that we multiply the independent term of these powers by the primitive function. We shall thus obtain the primitive function whose index is augmented by any number n by means of its differences.

Supposing that T and Z always have the preceding values, we shall have Z equal to the binomial $T - 1$; the product of V by the nth power of Z will then be equal to the product of V by the development of the nth power of the binomial $T - 1$. Repassing from the discriminant functions to their coefficients as has just been done, we shall have the nth difference of the primitive function expressed by the development of the nth power of the binomial $T - 1$, in which we substitute for the powers of T this same function whose index is augmented by the exponent of the power, and for the independent term of t, which is unity, the primitive function, which gives this difference by means of the consecutive terms of this function.

Placing δ before the primitive function expressing the derivative of this function, which multiplies the x power of t in the product of V by T, and Δ expressing the same derivative in the product of V by Z, we are led by that which precedes to this general result: whatever may be the function of the variable t represented by T and Z, we may, in the development of all the identical equations susceptible of being formed among these functions, substitute the characters δ and $\varDelta$ in place of T and Z, provided that we write the primitive function of the index in series with the powers and with the products of the powers of the characters, and that we multiply by this function the independent terms of these characters.

We are able by means of this general result to transform any certain power of a difference of the primitive function of the index x, in which x varies by unity, into a series of differences of the same function in which x varies by a certain number of units and reciprocally. Let us suppose that T be the i power of unity divided by $t - 1$, and that Z be always unity divided by $t - 1$; then the coefficient of the x power of t in the product of V by T will be the coefficient of the $x + i$ power of t in V less the coefficient of the x power of t; it will then be the finite difference of the primitive function of the index x in which we vary this index by the number i. It is easy to see that T is equal to the difference between the i power of the binomial $Z + 1$ and unity. The nth power of T is equal to the nth power of this difference. If in this equality we substitute in place of T and Z the characters δ and $\varDelta$, and after the development we place at the end of each term the primitive function of the index x, we shall have the nth difference of this function in which x varies by i units expressed by a series of differences of the same function in which x varies by unity. This series is

only a transformation of the difference which it
expresses and which is identical with it; but it is in
similar transformations that the power of analysis
resides.

The generality of analysis permits us to suppose in
this expression that n is negative. Then the negative
powers of δ and $\varDelta$ indicate the integrals. Indeed the
nth difference of the primitive function having for a
discriminant function the product of V by the nth power
of the binomial one divided by t less unity, the primitive function which is the nth integral of this difference
has for a discriminant function that of the same difference multiplied by the nth power taken less than the
binomial one divided by t minus one, a power to which
the same power of the character $\varDelta$ corresponds; this
power indicates then an integral of the same order, the
index x varying by unity; and the negative powers of
δ indicate equally the integrals x varying by i units.
We see, thus, in the clearest and simplest manner the
rationality of the analysis observed among the positive
powers and differences, and among the negative powers
and the integrals.

If the function indicated by δ placed before the
primitive function is zero, we shall have an equation of
finite differences, and V will be the discriminant function
of its integral. In order to obtain this discriminant
function we shall observe that in the product of V by
T all the powers of t ought to disappear except the
powers inferior to the order of the equation of differences; V is then equal to a fraction whose denominator
is T and whose numerator is a polynomial in which the
highest power of t is less by unity than the order of the

equation of differences.   The arbitrary coefficients of
the various powers of t in this polynomial, including
the power zero, will be determined by as many values
of the primitive function of the index when we make
successively x equal to zero, to one, to two, etc.
When the equation of differences is given we determine
T by putting all its terms in the first member and zero
in the second; by substituting in the first member unity
in place of the function which has the largest index;
the first power of t in place of the primitive function in
which this index is diminished by unity; the second
power of t for the primitive function where this index
is diminished by two units, and so on.   The coefficient
of the xth power of t in the development of the preceding expression of V will be the primitive function of x
or the integral of the equation of finite differences.
Analysis furnishes for this development various means,
among which we may choose that one which is most
suitable for the question proposed; this is an advantage
of this method of integration.

Let us conceive now that V be a function of the two
variables t and t' developed according to the powers
and products of these variables; the coefficient of any
product of the powers x and x' of t and t' will be a
function of the exponents or indices x and x' of these
powers; this function I shall call the *primitive function*
of which V is the discriminant function.

Let us multiply V by a function T of the two
variables t and t' developed like V in ratio of the
powers and the products of these variables; the product
will be the discriminant function of a derivative of the
primitive function; if T, for example, is equal to the

variable *t* plus the variable *t'* minus two, this derivative will be the primitive function of which we diminish by unity the index *x* plus this same primitive function of which we diminish by unity the index *x'* less two times the primitive function. Designating whatever *T* may be by the character *δ* placed before the primitive function, this derivative, the product of *V* by the *n*th power of *T*, will be the discriminant function of the derivative of the primitive function before which one places the *n*th power of the character *δ*. Hence result the theorems analogous to those which are relative to functions of a single variable.

Suppose the function indicated by the character *δ* be zero; one will have an equation of partial differences. If, for example, we make as before *T* equal to the variable *t* plus the variable *t'* — 2, we have zero equal to the primitive function of which we diminish by unity the index *x* plus the same function of which we diminish by unity the index *x'* minus two times the primitive function. The discriminant function *V* of the primitive function or of the integral of this equation ought then to be such that its product by *T* does not include at all the products of *t* by *t'*; but *V* may include separately the powers of *t* and those of *t'*, that is to say, an arbitrary function of *t* and an arbitrary function of *t'*; *V* is then a fraction whose numerator is the sum of these two arbitrary functions and whose denominator is *T*. The coefficient of the product of the *x*th power of *t* by the *x'* power of *t'* in the development of this fraction will then be the integral of the preceding equation of partial differences. This method of integrating this kind of equations seems to me the simplest and the easiest by the employment of the various analytical processes for the development of rational fractions.

More ample details in this matter would be scarcely understood without the aid of calculus.

Considering equations of infinitely small partial differences as equations of finite partial differences in which nothing is neglected, we are able to throw light upon the obscure points of their calculus, which have been the subject of great discussions among geometricians. It is thus that I have demonstrated the possibility of introducing discontinued functions in their integrals, provided that the discontinuity takes place only for the differentials of the order of these equations or of a superior order. The transcendent results of calculus are, like all the abstractions of the understanding, general signs whose true meaning may be ascertained only by repassing by metaphysical analysis to the elementary ideas which have led to them; this often presents great difficulties, for the human mind tries still less to transport itself into the future than to retire within itself. The comparison of infinitely small differences with finite differences is able similarly to shed great light upon the metaphysics of infinitesimal calculus.

It is easily proven that the finite nth difference of a function in which the increase of the variable is E being divided by the nth power of E, the quotient reduced in series by ratio to the powers of the increase E is formed by a first term independent of E. In the measure that E diminishes, the series approaches more and more this first term from which it can differ only by quantities less than any assignable magnitude.

This term is then the limit of the series and expresses in differential calculus the infinitely small *n*th difference of the function divided by the *n*th power of the infinitely small increase.

Considering from this point of view the infinitely small differences, we see that the various operations of differential calculus amount to comparing separately in the development of identical expressions the finite terms or those independent of the increments of the variables which are regarded as infinitely small; this is rigorously exact, these increments being indeterminant. Thus differential calculus has all the exactitude of other algebraic operations.

The same exactitude is found in the applications of differential calculus to geometry and mechanics. If we imagine a curve cut by a secant at two adjacent points, naming E the interval of the ordinates of these two points, E will be the increment of the abscissa from the first to the second ordinate. It is easy to see that the corresponding increment of the ordinate will be the product of E by the first ordinate divided by its subsecant; augmenting then in this equation of the curve the first ordinate by this increment, we shall have the equation relative to the second ordinate. The difference of these two equations will be a third equation which, developed by the ratio of the powers of E and divided by E, will have its first term independent of E, which will be the limit of this development. This term, equal to zero, will give then the limit of the subsecants, a limit which is evidently the subtangent.

This singularly happy method of obtaining the subtangent is due to Fermat, who has extended it to transcendent curves. This great geometrician expresses by the character E the increment of the abscissa; and considering only the first power of this increment, he determines exactly as we do by differential calculus the subtangents of the curves, their points of inflection, the *maxima* and *minima* of their ordinates, and in general those of rational functions. We see likewise by his beautiful solution of the problem of the refraction of light inserted in the *Collection of the Letters of Descartes* that he knows how to extend his methods to irrational functions in freeing them from irrationalities by the elevation of the roots to powers. Fermat should be regarded, then, as the true discoverer of Differential Calculus. Newton has since rendered this calculus more analytical in his *Method of Fluxions*, and simplified and generalized the processes by his beautiful theorem of the binomial. Finally, about the same time Leibnitz has enriched differential calculus by a notation which, by indicating the passage from the finite to the infinitely small, adds to the advantage of expressing the general results of calculus that of giving the first approximate values of the differences and of the sums of the quantities; this notation is adapted of itself to the calculus of partial differentials.

We are often led to expressions which contain so many terms and factors that the numerical substitutions are impracticable. This takes place in questions of probability when we consider a great number of events. Meanwhile it is necessary to have the numerical value of the formulæ in order to know with what probability the results are indicated, which the events develop by multiplication. It is necessary especially to have the law according to which this probability continually approaches certainty, which it will finally attain if the number of events were infinite. In order to obtain this law I considered that the definite integrals of differentials multiplied by the factors raised to great powers would give by integration the formulæ composed of a great number of terms and factors. This remark brought me to the idea of transforming into similar integrals the complicated expressions of analysis and the integrals of the equation of differences. I fulfilled this condition by a method which gives at the same time the function comprised under the integral sign and the limits of the integration. It offers this remarkable thing, that the function is the same discriminant function of the expressions and the proposed equations; this attaches this method to the theory of discriminant functions of which it is thus the complement. Further, it would only be a question of reducing the definite integral to a converging series. This I have obtained by a process which makes the series converge with as much more rapidity as the formula which it represents is more complicated, so that it is more exact as it becomes more necessary. Frequently the series has for a factor the square root of the ratio of the circumference to the diameter; sometimes it depends upon other transcendents whose number is infinite.

An important remark which pertains to great generality of analysis, and which permits us to extend this method to formulæ and to equations of difference which the theory of probability presents most frequently, is that the series to which one comes by supposing the limits of the definite integrals to be real and positive take place equally in the case where the equation which determines these limits has only negative or imaginary roots. These passages from the positive to the negative and from the real to the imaginary, of which I first have made use, have led me further to the values of many singular definite integrals, which I have accordingly demonstrated directly. We may then consider these passages as a means of discovery parallel to induction and analogy long employed by geometricians, at first with an extreme reserve, afterwards with entire confidence, since a great number of examples has justified its use. In the mean time it is always necessary to confirm by direct demonstrations the results obtained by these divers means.

I have named the ensemble of the preceding methods the *Calculus of Discriminant Functions;* this calculus serves as a basis for the work which I have published under the title of the *Analytical Theory of Probabilities.* It is connected with the simple idea of indicating the repeated multiplications of a quantity by itself or its entire and positive powers by writing toward the top of the letter which expresses it the numbers which mark the degrees of these powers.

This notation, employed by Descartes in his *Geometry* and generally adopted since the publication of this important work, is a little thing, especially when compared with the theory of curves and variable functions by which this great geometrician has established the foundations of modern calculus. But the language of analysis, most perfect of all, being in itself a powerful instrument of discoveries, its notations, especially when they are necessary and happily conceived, are so many germs of new calculi. This is rendered appreciable by this example.

Wallis, who in his work entitled *Arithmetica Infinitorum*, one of those which have most contributed to the progress of analysis, has interested himself especially in following the thread of induction and analogy, considered that if one divides the exponent of a letter by two, three, etc., the quotient will be accordingly the Cartesian notation, and when division is possible the exponent of the square, cube, etc., root of the quantity which represents the letter raised to the dividend exponent. Extending by analogy this result to the case where division is impossible, he considered a quantity raised to a fractional exponent as the root of the degree indicated by the denominator of this fraction—namely, of the quantity raised to a power indicated by the numerator. He observed then that, according to the Cartesian notation, the multiplication of two powers of the same letter amounts to adding their exponents, and that their division amounts to subtracting the exponents of the power of the divisor from that of the power of the dividend, when the second of these exponents is greater than the first. Wallis extended this result to the case where the first exponent is equal to or greater than the second, which makes the difference zero or negative. He supposed then that a negative exponent indicates unity divided by the quantity raised to the same exponent taken positively. These remarks led him to integrate generally the monomial differentials, whence he inferred the definite integrals of a particular kind of binomial differentials whose exponent is a positive integral

number.   The observation then of the law of the numbers which express these integrals, a series of inter-
polations and happy inductions where one perceives
the germ of the calculus of definite integrals which has
so much exercised geometricians and which is one of
the fundaments of my new *Theory of Probabilities*,
gave him the ratio of the area of the circle to the square
of its diameter expressed by an infinite product, which,
when one stops it, confines this ratio to limits more and
more converging; this is one of the most singular
results in analysis.   But it is remarkable that Wallis,
who had so well considered the fractional exponents
of radical powers, should have continued to note these
powers as had been done before him.   Newton in his
Letters to Oldembourg, if I am not mistaken, was the
first to employ the notation of these powers by fractional exponents.   Comparing by the way of induction,
of which Wallis had made such a beautiful use, the
exponents of the powers of the binomial with the
coefficients of the terms of its development in the case
where this exponent is integral and positive, he determined the law of these coefficients and extended it by
analogy to fractional and negative powers.   These
various results, based upon the notation of Descartes,
show his influence on the progress of analysis.   It has
still the advantage of giving the simplest and fairest
idea of logarithms, which are indeed only the exponents
of a magnitude whose successive powers, increasing by
infinitely small degrees, can represent all numbers.

But the most important extension that this notation
has received is that of variable exponents, which con-
stitutes exponential calculus, one of the most fruitful

branches of modern analysis. Leibnitz was the first to indicate the transcendents by variable exponents, and thereby he has completed the system of elements of which a finite function can be composed; for every finite explicit function of a variable may be reduced in the last analysis to simple magnitudes, combined by the method of addition, subtraction, multiplication, and division and raised to constant or variable powers. The roots of the equations formed from these elements are the implicit functions of the variable. It is thus that a variable has for a logarithm the exponent of the power which is equal to it in the series of the powers of the number whose hyperbolic logarithm is unity, and the logarithm of a variable of it is an implicit function.

Leibnitz thought to give to his differential character the same exponents as to magnitudes; but then in place of indicating the repeated multiplications of the same magnitude these exponents indicate the repeated differentiations of the same function. This new extension of the Cartesian notation led Leibnitz to the analogy of positive powers with the differentials, and the negative powers with the integrals. Lagrange has followed this singular analogy in all its developments; and by series of inductions which may be regarded as one of the most beautiful applications which have ever been made of the method of induction he has arrived at general formulæ which are as curious as useful on the transformations of differences and of integrals the ones into the others when the variables have divers finite increments and when these increments are infinitely small. But he has not given the demonstrations of it which appear to him difficult. The theory of discriminant functions extends the Cartesian notations to some of its characters; it shows with proof the analogy of the powers and operations indicated by these characters; so that it may still be regarded as the exponential calculus of characters.   All that concerns the series and the integration of equations of differences springs from it with an extreme facility.

PART II.

APPLICATIONS OF THE CALCULUS OF PROBABILITIES.

CHAPTER VI.

GAMES OF CHANCE.

THE combinations which games present were the object of the first investigations of probabilities. In an infinite variety of these combinations many of them lend themselves readily to calculus; others require more difficult calculi; and the difficulties increasing in the measure that the combinations become more complicated, the desire to surmount them and curiosity have excited geometricians to perfect more and more this kind of analysis. It has been seen already that the benefits of a lottery are easily determined by the theory of combinations. But it is more difficult to know in how many draws one can bet one against one, for example that all the numbers will be drawn, n being the number of numbers, r that of the numbers drawn at each draw, and i the unknown number of draws. The expression of the probability of drawing all the numbers depends upon the nth finite difference of the i power of a product of r consecutive numbers. When the number n is considerable the search for the value of i which renders this probability equal to $\frac{1}{2}$ becomes impossible at least unless this difference is converted into a very converging series. This is easily done by the method here below indicated by the approximations of functions of very large numbers. It is found thus since the lottery is composed of ten thousand numbers, one of which is drawn at each draw, that there is a disadvantage in betting one against one that all the numbers will be drawn in 95767 draws and an advantage in making the same bet for 95768 draws. In the lottery of France this bet is disadvantageous for 85 draws and advantageous for 86 draws.

Let us consider again two players, A and B, playing together at heads and tails in such a manner that at each throw if heads turns up A gives one counter to B, who gives him one if tails turns up; the number of counters of B is limited, while that of A is unlimited, and the game is to end only when B shall have no more counters. We ask in how many throws one should bet one to one that the game will end. The expression of the probability that the game will end in an i number of throws is given by a series which comprises a great number of terms and factors if the number of counters of B is considerable; the search for the value of the unknown i which renders this series $\frac{1}{2}$ would then be impossible if we did not reduce the same to a very convergent series. In applying to it the method of which we have just spoken, we find a very simple expression for the unknown from which it results that if, for example, B has a hundred counters, it is a bet of a little less than one against one that the game will end in 23780 throws, and a bet of a little more than one against one that it will end in 23781 throws.

These two examples added to those we have already given are sufficient to shows how the problems of games have contributed to the perfection of analysis.

CHAPTER VII.

CONCERNING THE UNKNOWN INEQUALITIES WHICH MAY EXIST AMONG CHANCES WHICH ARE SUPPOSED EQUAL.

INEQUALITIES of this kind have upon the results of the calculation of probabilities a sensible influence which deserves particular attention. Let us take the game of heads and tails, and let us suppose that it is equally easy to throw the one or the other side of the coin. Then the probability of throwing heads at the first throw is $\frac{1}{2}$ and that of throwing it twice in succession is $\frac{1}{4}$. But if there exist in the coin an inequality which causes one of the faces to appear rather than the other without knowing which side is favored by this inequality, the probability of throwing heads at the first throw will always be $\frac{1}{2}$; because of our ignorance of which face is favored by the inequality the probability of the simple event is increased if this inequality is favorable to it, just so much is it diminished if the inequality is contrary to it. But in this same ignorance the probability of throwing heads twice in succession is increased. Indeed this probability is that of throwing heads at the first throw multiplied by the probability that having thrown it at the first throw it will be thrown at the second; but its happening at the first throw is a reason for belief that the inequality of the coin favors it; the unknown inequality increases, then, the probability of throwing heads at the second throw; it consequently increases the product of these two probabilities. In order to submit this matter to calculus let us suppose that this inequality increases by a twentieth the probability of the simple event which it favors. If this event is heads, its probability will be $\frac{1}{2}$ plus $\frac{1}{20}$, or $\frac{11}{20}$, and the probability of throwing it twice in succession will be the square of $\frac{11}{20}$, or $\frac{121}{400}$. If the favored event is tails, the probability of heads, will be $\frac{1}{2}$ minus $\frac{1}{20}$, or $\frac{9}{20}$, and the probability of throwing it twice in succession will be $\frac{81}{400}$. Since we have at first no reason for believing that the inequality favors one of these events rather than the other, it is clear that in order to have the probability of the compound event heads heads it is necessary to add the two preceding probabilities and take the half of their sum, which gives $\frac{101}{400}$ for this probability, which exceeds $\frac{1}{4}$ by $\frac{1}{400}$ or by the square of the favor $\frac{1}{20}$ that the inequality adds to the possibilities of the event which it favors. The probability of throwing tails tails is similarly $\frac{101}{400}$, but the probability of throwing heads tails or tails heads is each $\frac{99}{400}$; for the sum of these four probabilities ought to equal certainty or unity. We find thus generally that the constant and unknown causes which favor simple events which are judged equally possible always increase the probability of the repetition of the same simple event.

In an even number of throws heads and tails ought

both to happen either an even number of times or odd number of times. The probability of each of these cases is $\frac{1}{2}$ if the possibilities of the two faces are equal; but if there is between them an unknown inequality, this inequality is always favorable to the first case.

Two players whose skill is supposed to be equal play on the conditions that at each throw that one who loses gives a counter to his adversary, and that the game continues until one of the players has no more counters. The calculation of the probabilities shows us that for the equality of the play the throws of the players ought to be an inverse ratio to their counters. But if there is between the players a small unknown inequality, it favors that one of the players who has the smallest number of counters. His probability of winning the game increases if the players agree to double or triple their counters; and it will be $\frac{1}{2}$ or the same as the probability of the other player in the case where the number of their counters should become infinite, preserving always the same ratio.

One may correct the influence of these unknown inequalities by submitting them themselves to the chances of hazard. Thus at the play of heads and tails, if one has a second coin which is thrown each time with the first and one agrees to name constantly heads the face turned up by the second coin, the probability of throwing heads twice in succession with the first coin will approach much nearer $\frac{1}{4}$ than in the case of a single coin. In this last case the difference is the square of the small increment of possibility that the unknown inequality gives to the face of the first coin which it favors; in the other case this difference is the quadruple product of this square by the corresponding square relative to the second coin.

Let there be thrown into an urn a hundred numbers from 1 to 100 in the order of numeration, and after having shaken the urn in order to mix the numbers one is drawn; it is clear that if the mixing has been well done the probabilities of the drawing of the numbers will be the same. But if we fear that there is among them small differences dependent upon the order according to which the numbers have been thrown into the urn, we shall diminish considerably these differences by throwing into a second urn the numbers according to the order of their drawing from the first urn, and by shaking then this second urn in order to mix the numbers. A third urn, a fourth urn, etc., would diminish more and more these differences already inappreciable in the second urn.

CHAPTER VIII.

CONCERNING THE LAWS OF PROBABILITY WHICH RESULT FROM THE INDEFINITE MULTIPLICATION OF EVENTS.

AMID the variable and unknown causes which we comprehend under the name of *chance*, and which render uncertain and irregular the march of events, we see appearing, in the measure that they multiply, a striking regularity which seems to hold to a design and which has been considered as a proof of Providence. But in reflecting upon this we soon recognize that this regularity is only the development of the respective possibilities of simple events which ought to present themselves more often when they are more probable. Let us imagine, for example, an urn which contains white balls and black balls; and let us suppose that each time a ball is drawn it is put back into the urn before proceeding to a new draw. The ratio of the number of the white balls drawn to the number of black balls drawn will be most often very irregular in the first drawings; but the variable causes of this irregularity produce effects alternately favorable and unfavorable to the regular march of events which destroy each other mutually in the totality of a great number of draws, allowing us to perceive more and more the ratio of white balls to the black balls contained in the urn, or the respective possibilities of drawing a white ball or black ball at each draw. From this results the following theorem.

The probability that the ratio of the number of white balls drawn to the total number of balls drawn does not deviate beyond a given interval from the ratio of the number of white balls to the total number of balls contained in the urn, approaches indefinitely to certainty by the indefinite multiplication of events, however small this interval.

This theorem indicated by common sense was difficult to demonstrate by analysis. Accordingly the illustrious geometrician Jacques Bernouilli, who first has occupied himself with it, attaches great importance to the demonstrations he has given. The calculus of discriminant functions applied to this matter not only demonstrates with facility this theorem, but still more it gives the probability that the ratio of the events observed deviates only in certain limits from the true ratio of their respective possibilities.

One may draw from the preceding theorem this consequence which ought to be regarded as a general law, namely, that the ratios of the acts of nature are very nearly constant when these acts are considered in great number. Thus in spite of the variety of years the sum of the productions during a considerable number of years is sensibly the same; so that man by useful foresight is able to provide against the irregularity of the seasons by spreading out equally over all the seasons the goods which nature distributes in an unequal manner. I do not except from the above law results due to moral causes. The ratio of annual births to the population, and that of marriages to births, show only small variations; at Paris the number of annual births is almost the same, and I have heard it said at the post-office in ordinary seasons the number of letters thrown aside on account of defective addresses changes little each year; this has likewise been observed at London.

It follows again from this theorem that in a series of events indefinitely prolonged the action of regular and constant causes ought to prevail in the long run over that of irregular causes. It is this which renders the gains of the lotteries just as certain as the products of agriculture; the chances which they reserve assure them a benefit in the totality of a great number of throws. Thus favorable and numerous chances being constantly attached to the observation of the eternal principles of reason, of justice, and of humanity which establish and maintain societies, there is a great advantage in conforming to these principles and of grave inconvenience in departing from them. If one consult histories and his own experience, one will see all the facts come to the aid of this result of calculus. Consider the happy effects of institutions founded upon reason and the natural rights of man among the peoples who have known how to establish and preserve them. Consider again the advantages which good faith has procured for the governments who have made it the basis of their conduct and how they have been indemnified for the sacrifices which a scrupulous exactitude in keeping their engagements has cost them. What immense credit at home! What preponderance abroad! On the contrary, look into what an abyss of misfortunes nations have often been precipitated by the ambition and the perfidy of their chiefs. Every time that a great power intoxicated by the love of conquest aspires to universal domination the sentiment of independence produces among the menaced nations a coalition of which it becomes almost always the victim. Similarly in the midst of the variable causes which extend or restrain the divers states, the natural limits acting as constant causes ought to end by prevailing. It is important then to the stability as well as to the happiness of empires not to extend them beyond those limits into which they are led again without cessation by the action of the causes; just as the waters of the seas raised by violent tempests fall again into their basins by the force of gravity. It is again a result of the calculus of probabilities confirmed by numerous and melancholy experiences. History treated from the point of view of the influence of constant causes would unite to the interest of curiosity that of offering to man most useful lessons. Sometimes we attribute the inevitable results of these causes to the accidental circumstances which have produced their action. It is, for example, against the nature of things that one people should ever be governed by another when a vast sea or a great distance separates them. It may be affirmed that in the long run this constant cause, joining itself without ceasing to the variable causes which act in the same way and which the course of time develops, will end by finding them sufficiently strong to give to a subjugated people its natural independence or to unite it to a powerful state which may be contiguous.

In a great number of cases, and these are the most important of the analysis of hazards, the possibilities of simple events are unknown and we are forced to search in past events for the indices which can guide us in our conjectures about the causes upon which they depend. In applying the analysis of discriminant functions to the principle elucidated above on the probability of the causes drawn from the events observed, we are led to the following theorem.

When a simple event or one composed of several simple events, as, for instance, in a game, has been repeated a great number of times the possibilities of the simple events which render most probable that which has been observed are those that observation indicates with the greatest probability; in the measure that the observed event is repeated this probability increases and would end by amounting to certainty if the numbers of repetitions should become infinite.

There are two kinds of approximations: the one is relative to the limits taken on all sides of the possibilities which give to the past the greatest probability; the other approximation is related to the probability that these possibilities fall within these limits. The repetition of the compound event increases more and more this probability, the limits remaining the same; it reduces more and more the interval of these limits, the probability remaining the same; in infinity this interval becomes zero and the probability changes to certainty.

If we apply this theorem to the ratio of the births of

boys to that of girls observed in the different countries
of Europe, we find that this ratio, which is everywhere
about equal to that of 22 to 21, indicates with an
extreme probability a greater facility in the birth of
boys. Considering further that it is the same at Naples
and at St. Petersburg, we shall see that in this regard
the influence of climate is without effect. We might
then suspect, contrary to the common belief, that this
predominance of masculine births exists even in the
Orient. I have consequently invited the French
scholars sent to Egypt to occupy themselves with this
interesting question; but the difficulty in obtaining
exact information about the births has not permitted
them to solve it. Happily, M. de Humboldt has not
neglected this matter among the innumerable new
things which he has observed and collected in America
with so much sagacity, constancy, and courage. He
has found in the tropics the same ratio of the births as
we observe in Paris; this ought to make us regard the
greater number of masculine births as a general law of
the human race. The laws which the different kinds
of animals follow in this regard seem to me worthy of
the attention of naturalists.

The fact that the ratio of births of boys to that of
girls differs very little from unity even in the great
number of the births observed in a place would offer in
this regard a result contrary to the general law, without
which we should be right in concluding that this law
did not exist. In order to arrive at this result it is
necessary to employ great numbers and to be sure that
it is indicated by great probability. Buffon cites, for
example, in his *Political Arithmetic* several communi-

ties of Bourgogne where the births of girls have surpassed those of boys.   Among these communities that
of Carcelle-le-Grignon presents in 2009 births during
five years 1026 girls and 983 boys.   Although these
numbers are considerable, they indicate, however, only
a greater possibility in the births of girls with a probability of $\frac{9}{10}$, and this probability, smaller than that of
not throwing heads four times in succession in the game
of heads and tails, is not sufficient to investigate the
cause for this anomaly, which, according to all probability, would disappear if one should follow during a
century the births in this community.

The registers of births, which are kept with care in
order to assure the condition of the citizens, may serve
in determining the population of a great empire without
recurring to the enumeration of its inhabitants—a
laborious operation and one difficult to make with
exactitude.   But for this it is necessary to know the
ratio of the population to the annual births.   The most
precise means of obtaining it consists, first, in choosing
in the empire districts distributed in an almost equal
manner over its whole surface, so as to render the
general result independent of local circumstances;
second, in enumerating with care for a given epoch the
inhabitants of several communities in each of these districts; third, by determining from the statement of the
births during several years which precede and follow
this epoch the mean number corresponding to the
annual births.   This number, divided by that of the
inhabitants, will give the ratio of the annual births to
the population in a manner more and more accurate
as the enumeration becomes more considerable.   The

government, convinced of the utility of a similar enumeration, has decided at my request to order its execution. In thirty districts spread out equally over the whole of France, communities have been chosen which would be able to furnish the most exact information. Their enumerations have given 2037615 individuals as the total number of their inhabitants on the 23d of September, 1802. The statement of the births in these communities during the years 1800, 1801, and 1802 have given:

Births.	Marriages.	Deaths.
110312 boys	46037	103659 men
105287 girls		99443 women

The ratio of the population to annual births is then $28\frac{352845}{1000000}$; it is greater than had been estimated up to this time. Multiplying the number of annual births in France by this ratio, we shall have the population of this kingdom. But what is the probability that the population thus determined will not deviate from the true population beyond a given limit? Resolving this problem and applying to its solution the preceding data, I have found that, the number of annual births in France being supposed to be 1000000, which brings the population to 28352845 inhabitants, it is a bet of almost 300000 against 1 that the error of this result is not half a million.

The ratio of the births of boys to that of girls which the preceding statement offers is that of 22 to 21; and the marriages are to the births as 3 is to 4.

At Paris the baptisms of children of both sexes vary a little from the ratio of 22 to 21. Since 1745, the epoch in which one has commenced to distinguish the sexes upon the birth-registers, up to the end of 1784, there have been baptized in this capital 393386 boys and 377555 girls. The ratio of the two numbers is almost that of 25 to 24; it appears then at Paris that a particular cause approximates an equality of baptisms of the two sexes. If we apply to this matter the calculus of probabilities, we find that it is a bet of 238 to 1 in favor of the existence of this cause, which is sufficient to authorize the investigation. Upon reflection it has appeared to me that the difference observed holds to this, that the parents in the country and the provinces, finding some advantage in keeping the boys at home, have sent to the Hospital for Foundlings in Paris fewer of them relative to the number of girls according to the ratio of births of the two sexes. This is proved by the statement of the registers of this hospital. From the beginning of 1745 to the end of 1809 there were entered 163499 boys and 159405 girls. The first of these numbers exceeds only by $\frac{1}{38}$ the second, which it ought to have surpassed at least by $\frac{1}{24}$. This confirms the existence of the assigned cause, namely, that the ratio of births of boys to those of girls is at Paris that of 22 to 21, no attention having been paid to foundlings.

The preceding results suppose that we may compare the births to the drawings of balls from an urn which contains an infinite number of white balls and black balls so mixed that at each draw the chances of drawing ought to be the same for each ball; but it is possible that the variations of the same seasons in different years may have some influence upon the annual ratio of the births of boys to those of girls. The Bureau of Longitudes of France publishes each year in its annual the tables of the annual movement of the population of the kingdom. The tables already published commence in 1817; in that year and in the five following years there were born 2962361 boys and 2781997 girls, which gives about $\frac{16}{15}$ for the ratio of the births of boys to that of girls. The ratios of each year vary little from this mean result; the smallest ratio is that of 1822, where it was only $\frac{17}{16}$; the greatest is of the year 1817, when it was $\frac{15}{14}$. These ratios vary appreciably from the ratio of $\frac{22}{21}$ found above. Applying to this deviation the analysis of probabilities in the hypothesis of the comparison of births to the drawings of balls from an urn, we find that it would be scarcely probable. It appears, then, to indicate that this hypothesis, although closely approximated, is not rigorously exact. In the number of births which we have just stated there are of natural children 200494 boys and 190698 girls. The ratio of masculine and feminine births was then in this regard $\frac{20}{19}$, smaller than the mean ratio of $\frac{16}{15}$. This result is in the same sense as that of the births of foundlings; and it seems to prove that in the class of natural children the births of the two sexes approach more nearly equality than in the class of legitimate children. The difference of the climates from the north to the south of France does not appear to influence appreciably the ratio of the births of boys and girls. The thirty most southern districts have given $\frac{16}{15}$ for this ratio, the same as that of entire France.

The constancy of the superiority of the births of boys over girls at Paris and at London since they have been observed has appeared to some scholars to be a proof of Providence, without which they have thought that the irregular causes which disturb without ceasing the course of events ought several times to have rendered the annual births of girls superior to those of boys.

But this proof is a new example of the abuse which has been so often made of final causes which always disappear on a searching examination of the questions when we have the necessary data to solve them. The constancy in question is a result of regular causes which give the superiority to the births of boys and which extend it to the anomalies due to hazard when the number of annual births is considerable. The investigation of the probability that this constancy will maintain itself for a long time belongs to that branch of the analysis of hazards which passes from past events to the probability of future events; and taking as a basis the births observed from 1745 to 1784, it is a bet of almost 4 against 1 that at Paris the annual births of boys will constantly surpass for a century the births of girls; there is then no reason to be astonished that this has taken place for a half-century.

Let us take another example of the development of constant ratios which events present in the measure that they are multiplied. Let us imagine a series of urns arranged circularly, and each containing a very great number of white balls and black balls; the ratio of white balls to the black in the urns being originally very different and such, for example, that one of these urns contains only white balls, while another contains only black balls. If one draws a ball from the first urn in order to put it into the second, and, after having

shaken the second urn in order to mix well the new
ball with the others, one draws a ball to put it into the
third urn, and so on to the last urn, from which is drawn
a ball to put into the first, and if this series is recom-
menced continually, the analysis of probability shows
us that the ratios of the white balls to the black in these
urns will end by being the same and equal to the ratio
of the sum of all the white balls to the sum of all the
black balls contained in the urns. Thus by this regular
mode of change the primitive irregularity of these ratios
disappears eventually in order to make room for the
most simple order. Now if among these urns one
intercalate new ones in which the ratio of the sum of
the white balls to the sum of the black balls which they
contain differs from the preceding, continuing indefinitely in the totality of the urns the drawings which we
have just indicated, the simple order established in the
old urns will be at first disturbed, and the ratios of the
white balls to the black balls will become irregular;
but little by little this irregularity will disappear in
order to make room for a new order, which will finally
be that of the equality of the ratios of the white balls
to the black balls contained in the urns. We may
apply these results to all the combinations of nature in
which the constant forces by which their elements are
animated establish regular modes of action, suited to
bring about in the very heart of chaos systems governed
by admirable laws.

The phenomena which seem the most dependent
upon hazard present, then, when multiplied a tendency
to approach without ceasing fixed ratios, in such a
manner that if we conceive on all sides of each of these

ratios an interval as small as desired, the probability that the mean result of the observations falls within this interval will end by differing from certainty only by a quantity greater than an assignable magnitude.   Thus by the calculations of probabilities applied to a great number of observations we may recognize the existence of these ratios.   But before seeking the causes it is necessary, in order not to be led into vain speculations, to assure ourselves that they are indicated by a probability which does not permit us to regard them as anomalies due to hazard.   The theory of discriminant functions gives a very simple expression for this probability, which is obtained by integrating the product of the differential of the quantity of which the result deduced from a great number of observations varies from the truth by a constant less than unity, dependent upon the nature of the problem, and raised to a power whose exponent is the ratio of the square of this variation to the number of observations.   The integral taken between the limits given and divided by the same integral, applied to a positive and negative infinity, will express the probability that the variation from the truth is comprised between these limits.   Such is the general law of the probability of results indicated by a great number of observations.

CHAPTER IX.

THE APPLICATION OF THE CALCULUS OF PROB-ABILITIES TO NATURAL PHILOSOPHY.

THE phenomena of nature are most often enveloped by so many strange circumstances, and so great a number of disturbing causes mix their influence, that it is very difficult to recognize them. We may arrive at them only by multiplying the observations or the experiences, so that the strange effects finally destroy reciprocally each other, the mean results putting in evidence those phenomena and their divers elements. The more numerous the number of observations and the less they vary among themselves the more their results approach the truth. We fulfil this last condition by the choice of the methods of observations, by the precision of the instruments, and by the care which we take to observe closely; then we determine by the theory of probabilities the most advantageous mean results or those which give the least value of the error. But that is not sufficient; it is further necessary to appreciate the probability that the errors of these results are comprised in the given limits; and without this we have only an imperfect knowledge of the degree of exactitude obtained. Formulæ suitable to these matters are then true improvements of the method of sciences, and it is indeed important to add them to this method.   The analysis which they require is the most delicate and the most difficult of the theory of probabilities; it is one of the principal objects of the work which I have published upon this theory, and in which I have arrived at formulæ of this kind which have the remarkable advantage of being independent of the law of the probability of errors and of including only the quantities given by the observations themselves and their expressions.

Each observation has for an analytic expression a function of the elements which we wish to determine; and if these elements are nearly known, this function becomes a linear function of their corrections.   In equating it to the observation itself there is formed *an equation of condition*.   If we have a great number of similar equations, we combine them in such a manner as to obtain as many final equations as there are elements whose corrections we determine then by resolving these equations.   But what is the most advantageous manner of combining equations of condition in order to obtain final equations?   What is the law of the probabilities of errors of which the elements are still susceptible that we draw from them?   This is made clear to us by the theory of probabilities.   The formation of a final equation by means of the equation of condition amounts to multiplying each one of these by an indeterminate factor and by uniting the products; it is necessary to choose the system of factors which gives the smallest opportunity for error.   But it is apparent that if we multiply the possible errors of an element by their respective probabilities, the most advantageous system will be that in which the sum of these products all, taken, positively is a *minimum;* for a positive or a negative error ought to be considered as a loss. Forming, then, this sum of products, the condition of the *minimum* will determine the system of factors which it is expedient to adopt, or the most advantageous system. We find thus that this system is that of the coefficients of the elements in each equation of condition; so that we form a first final equation by multiplying respectively each equation of condition by its coefficient of the first element and by uniting all these equations thus multiplied. We form a second final equation by employing in the same manner the coefficients of the second element, and so on. In this manner the elements and the laws of the phenomena obtained in the collection of a great number of observations are developed with the most evidence.

The probability of the errors which each element still leaves to be feared is proportional to the number whose hyperbolic logarithm is unity raised to a power equal to the square of the error taken as a minus quantity and multiplied by a constant coefficient which may be considered as the modulus of the probability of the errors; because, the error remaining the same, its probability decreases with rapidity when the former increases; so that the element obtained weighs, if I may thus speak toward the truth, as much more as this modulus is greater. I would call for this reason this modulus the *weight* of the element or of the result. This weight is the greatest possible in the system of factors—the most advantageous; it is this which gives to this system superiority over others. By a remarkable analogy of this weight with those of bodies compared at their common centre of gravity it results that if the same element is given by divers systems, composed each of a great number of observations, the most advantageous, the mean result of their totality is the sum of the products of each partial result by its weight. Moreover, the total weight of the results of the divers systems is the sum of their partial weights; so that the probability of the errors of the mean result of their totality is proportional to the number which has unity for an hyperbolic logarithm raised to a power equal to the square of the error taken as minus and multiplied by the sum of the weights. Each weight depends in truth upon the law of the probability of error of each system, and almost always this law is unknown; but happily I have been able to eliminate the factor which contains it by means of the sum of the squares of the variations of the observations in this system from their mean result. It would then be desirable in order to complete our knowledge of the results obtained by the totality of a great number of observations that we write by the side of each result the weight which corresponds to it; analysis furnishes for this object both general and simple methods. When we have thus obtained the exponential which represents the law of the probability of errors, we shall have the probability that the error of the result is included within given limits by taking within the limits the integral of the product of this exponential by the differential of the error and multiplying it by the square root of the weight of the result divided by the circumference whose diameter is unity. Hence it follows that for the same probability the errors of the results are reciprocal to the square roots of their weights, which serves to compare their respective precision.

In order to apply this method with success it is necessary to vary the circumstances of the observations or the experiences in such a manner as to avoid the constant causes of error. It is necessary that the observations should be numerous, and that they should be so much the more so as there are more elements to determine; for the weight of the mean result increases as the number of observations divided by the number of the elements. It is still necessary that the elements follow in these observations a different course; for if the course of the two elements were exactly the same, which would render their coefficients proportional in equation of conditions, these elements would form only a single unknown quantity and it would be impossible to distinguish them by these observations. Finally it is necessary that the observations should be precise; this condition, the first of all, increases greatly the weight of the result the expression of which has for a divisor the sum of the squares of the deviations of the observations from this result. With these precautions we shall be able to make use of the preceding method and measure the degree of confidence which the results deduced from a great number of observations merit.

The rule which we have just given to conclude equations of condition, final equations, amount to rendering a minimum the sum of the squares of the errors of observations; for each equation of condition becomes exact by substituting in it the observation plus its error; and if we draw from it the expression of this error, it is easy to see that the condition of the *minimum* of the sum of the squares of these expressions gives the rule in question. This rule is the more precise as the observations are more numerous; but even in the case where their number is small it appears natural to employ the same rule which in all cases offers a simple means of obtaining without groping the corrections which we seek to determine. It serves further to compare the precision of the divers astronomical tables of the same star. These tables may always be supposed as reduced to the same form, and then they differ only by the epochs, the mean movements and the coefficients of the arguments; for if one of them contains a coefficient which is not found in the others, it is clear that this amounts to supposing zero in them as the coefficient of this argument. If now we rectify these tables by the totality of the good observations, they would satisfy the condition that the sum of the squares of the errors should be a minimum; the tables which, compared to a considerable number of observations, approach nearest this condition merit then the preference.

It is principally in astronomy that the method explained above may be employed with advantage. The astronomical tables owe the truly astonishing exactitude which they have attained to the precision of observations and of theories, and to the use of equations of conditions which cause to concur a great number of excellent observations in the correction of the same element. But it remains to determine the probability of the errors that this correction leaves still to be feared; and the method which I have just explained enables us to recognize the probability of these errors. In order to give some interesting applications of it I have profited by the immense work which M. Bouvard has just finished on the movements of Jupiter and Saturn, of which he has formed very precise tables. He has discussed with the greatest care the oppositions and quadratures of these two planets observed by Bradley and by the astronomers who have followed him down to the last years; he has concluded the corrections of the elements of their movement and their masses compared to that of the sun taken as unity. His calculations give him the mass of Saturn equal to the 3512th part of that of the sun. Applying to them my formulæ of probability, I find that it is a bet of 11,000 against one that the error of this result is not $\frac{1}{100}$ of its value, or that which amounts to almost the same—that after a century of new observations added to the preceding ones, and examined in the same manner, the new result will not differ by $\frac{1}{100}$ from that of M. Bouvard. This wise astronomer finds again the mass of Jupiter equal to the 1071th part of the sun; and my method of probability gives a bet of 1,000,000 to one that this result is not $\frac{1}{100}$ in error.

This method may be employed again with success in geodetic operations. We determine the length of the great arc on the surface of the earth by triangulation, which depends upon a base measured with exactitude. But whatever precision may be brought to the measure of the angles, the inevitable errors can, by accumulating, cause the value of the arc concluded from a great number of triangles to deviate appreciably from the truth.   We recognize this value, then, only imperfectly unless the probability that its error is comprised within given limits can be assigned.   The error of a geodetic result is a function of the errors of the angles of each triangle.   I have given in the work cited general formulæ in order to obtain the probability of the values of one or of several linear functions of a great number of partial errors of which we know the law of probability; we may then by means of these formulæ determine the probability that the error of a geodetic result is contained within the assigned limits, whatever may be the law of the probability of partial errors.   It is moreover more necessary to render ourselves independent of the law, since the most simple laws themselves are always infinitely less probable, seeing the infinite number of those which may exist in nature.   But the unknown law of partial errors introduces into the formulæ an indeterminant which does not permit of reducing them to numbers unless we are able to eliminate it.   We have seen that in astronomical questions, where each observation furnishes an equation of condition for obtaining the elements, we eliminate this determinant by means of the sum of the squares of the remainders when the most probable values of the elements have been substituted in each equation.   Geodetic questions not offering similar equations, it is necessary to seek another means of elimination.   The quantity by which the sum of the angles of each observed triangle surpasses two right angles plus the spherical excess furnishes this means.   Thus we replace by the sum of the squares of these quantities the sum of the squares of the remainders of the equations of condition;

and we may assign in numbers the probability that the error of the final result of a series of geodetic operations will not exceed a given quantity. But what is the most advantageous manner of dividing among the three angles of each triangle the observed sum of their errors? The analysis of probabilities renders it apparent that each angle ought to be diminished by a third of this sum, provided that the weight of a geodetic result be the greatest possible, which renders the same error less probable. There is then a great advantage in observing the three angles of each triangle and of correcting them as we have just said. Simple common sense indicates this advantage; but the calculation of probabilities alone is able to appreciate it and to render apparent that by this correction it becomes the greatest possible.

In order to assure oneself of the exactitude of the value of a great arc which rests upon a base measured at one of its extremities one measures a second base toward the other extremity; and one concludes from one of these bases the length of the other. If this length varies very little from the observation, there is all reason to believe that the chain of triangles which unites these bases is very nearly exact and likewise the value of the large arc which results from it. One corrects, then, this value by modifying the angles of the triangles in such a manner that the base is calculated according to the bases measured. But this may be done in an infinity of ways, among which is preferred that of which the geodetic result has the greatest weight, inasmuch as the same error becomes less probable. The analysis of probabilities gives formulæ for obtaining directly the most advantageous correction which results from the measurements of the several bases and the laws of probability which the multiplicity of the bases makes—laws which become very rapidly decreasing by this multiplicity.

Generally the errors of the results deduced from a great number of observations are the linear functions of the partial errors of each observation. The coefficients of these functions depend upon the nature of the problem and upon the process followed in order to obtain the results. The most advantageous process is evidently that in which the same error in the results is less probable than according to any other process. The application of the calculus of probabilities to natural philosophy consists, then, in determining analytically the probability of the values of these functions and in choosing their indeterminant coefficients in such a manner that the law of this probability should be most rapidly descending. Eliminating, then, from the formulæ by the data of the question the factor which is introduced by the almost always unknown law of the probability of partial errors, we may be able to evaluate numerically the probability that the errors of the results do not exceed a given quantity. We shall thus have all that may be desired touching the results deduced from a great number of observations.

Very approximate results may be obtained by other considerations. Suppose, for example, that one has a thousand and one observations of the same quantity; the arithmetical mean of all these observations is the result given by the most advantageous method. But one would be able to choose the result according to the condition that the sum of the variations from each partial value all taken positively should be a *minimum*. It appears indeed natural to regard as very approximate the result which satisfies this condition. It is easy to see that if one disposes the values given by the observations according to the order of magnitude, the value which will occupy the mean will fulfil the preceding condition, and calculus renders it apparent that in the case of an infinite number of observations it would coincide with the truth; but the result given by the most advantageous method is still preferable.

We see by that which precedes that the theory of probabilities leaves nothing arbitrary in the manner of distributing the errors of the observations; it gives for this distribution the most advantageous formulæ which diminishes as much as possible the errors to be feared in the results.

The consideration of probabilities can serve to distinguish the small irregularities of the celestial movements enveloped in the errors of observations, and to repass to the cause of the anomalies observed in these movements.

In comparing all the observations it was Ticho-Brahé who recognized the necessity of applying to the moon an equation of time different from that which had been applied to the sun and to the planets. It was similarly the totality of a great number of observations which made Mayer recognize that the coefficient of the inequality of the precession ought to be diminished a little for the moon. But since this diminution, although confirmed and even augmented by Mason, did not appear to result from universal gravitation, the majority of astronomers neglect it in their calculations. Having submitted to the calculation of probabilities a considerable number of lunar observations chosen for this purpose and which M. Bouvard consented to examine at my request, it appeared to me to be indicated with so strong a probability that I believed the cause of it ought to be investigated. I soon saw that it would be only the ellipticity of the terrestrial spheroid, neglected up to that time in the theory of the lunar movement as being able to produce only imperceptible terms. I concluded that these terms became perceptible by the successive integrations of differential equations. I determined then those terms by a particular analysis, and I discovered first the inequality of the lunar movement in latitude which is proportional to the sine of the longitude of the moon, which no astronomer before had suspected. I recognized then by means of this inequality that another exists in the lunar movement in longitude which produces the diminution observed by Mayer in the equation of the precession applicable to the moon. The quantity of this diminution and the coefficient of the preceding inequality in latitude are very appropriate to fix the oblateness of the earth. Having communicated my researches to M. Burg, who was occupied at that time in perfecting the tables of the moon by the comparison of all the good observations, I requested him to determine with a particular care these two quantities. By a very remarkable agreement the values which he has found give to the earth the same oblateness, $\frac{1}{305}$, which differs little from the mean derived from the measurements of the degrees of the meridian and the pendulum; but those regarded from the point of view of the influence of the errors of the observations and of the perturbing causes in these measurements, did not appear to me exactly determined by these lunar inequalities.

It was again by the consideration of probabilities that I recognized the cause of the secular equation of the moon. The modern observations of this star compared to the ancient eclipses had indicated to astronomers an acceleration in the lunar movement; but the geometricians, and particularly Lagrange, having vainly sought in the perturbations which this movement experienced the terms upon which this acceleration depends, reject it. An attentive examination of the ancient and modern observations and of the intermediary eclipses observed by the Arabians convinced me that it was indicated with a great probability. I took up again then from this point of view the lunar theory, and I recognized that the secular equation of the moon is due to the action of the sun upon this satellite, combined with the secular variation of the eccentricity of the terrestrial orb; this brought me to the discovery of the secular equations of the movements of the nodes and of the perigees of the lunar orbit, which equations had not been even suspected by astronomers. The very remarkable agreement of this theory with all the ancient and modern observations has brought it to a very high degree of evidence.

The calculus of probabilities has led me similarly to the cause of the great irregularities of Jupiter and Saturn. Comparing modern observations with ancient, Halley found an acceleration in the movement of Jupiter and a retardation in that of Saturn. In order to conciliate the observations he reduced the movements to two secular equations of contrary signs and increasing as the squares of the times passed since 1700. Euler and Lagrange submitted to analysis the alterations which the mutual attraction of these two planets ought to produce in these movements. They found in doing this the secular equations; but their results were so different that one of the two at least ought to be erroneous. I determined then to take up again this important problem of *celestial mechanics*, and I recognized the invariability of the mean planetary movements, which nullified the secular equations introduced by Halley in the tables of Jupiter and Saturn. Thus there remain, in order to explain the great irregularity of these planets, only the attractions of the comets to which many astronomers had effective recourse, or the existence of an irregularity over a long period produced in the movements of the two planets by their reciprocal action and affected by contrary signs for each of them. A theorem which I found in regard to the inequalities of this kind rendered this inequality very probable. According to this theorem, if the movement of Jupiter is accelerated, that of Saturn is retarded, which has already conformed to what Halley had noticed; moreover, the acceleration of Jupiter resulting from the same theorem is to the retardation of Saturn very nearly in the ratio of the secular equations proposed by Halley. Considering the mean movements of Jupiter and Saturn I was enabled easily to recognize that two times that of Jupiter differed only by a very small quantity from five times that of Saturn. The period of an irregularity which would have for an argument this difference would be about nine centuries. Indeed its coefficient would be of the order of the cubes of the eccentricities of the orbits; but I knew that by virtue of successive integrations it acquired for divisor the square of the very small multiplier of the time in the argument of this inequality which is able to give it a great value; the existence of this inequality appeared to me then very probable. The following observation increased then its probability. Supposing its argument zero toward the epoch of the observations of Ticho-Brahé, I saw that Halley ought to have found by the comparison of modern with ancient observations the alterations which he had indicated; while the comparison of the modern observations among themselves ought to offer contrary alterations similar to those which Lambert had concluded from this comparison. I did not then hesitate at all to undertake this long and tedious calculation necessary to assure myself of this inequality. It was entirely confirmed by the result of this calculation, which moreover made me recognize a great number of other inequalities of which the totality has inclined the tables of Jupiter and Saturn to the precision of the same observations.

It was again by means of the calculus of probabilities that I recognized the remarkable law of the mean movements of the three first satellites of Jupiter, according to which the mean longitude of the first minus three times that of the second plus two times that of the third is rigorously equal to the half-circumference. The approximation with which the mean movements of these stars satisfy this law since their discovery indicates its existence with an extreme probability. I sought then the cause of it in their mutual action. The searching examination of this action convinced me that it was sufficient if in the beginning the ratios of their mean movements had approached this law within certain limits, because their mutual action had established and maintained it rigorously. Thus these three bodies will balance one another eternally in space according to the preceding law unless strange causes, such as comets, should change suddenly their movements about Jupiter.

Accordingly it is seen how necessary it is to be attentive to the indications of nature when they are the result of a great number of observations, although in other respects they may be inexplicable by known means. The extreme difficulty of problems relative to the system of the world has forced geometricians to recur to the approximation which always leaves room for the fear that the quantities neglected may have an appreciable influence. When they have been warned of this influence by the observations, they have recurred to their analysis; in rectifying it they have always found the cause of the anomalies observed; they have determined the laws and often they have anticipated the observations in discovering the inequalities which it had not yet indicated. Thus one may say that nature itself has concurred in the analytical perfection of the theories based upon the principle of universal gravity; and this is to my mind one of the strongest proofs of the truth of this admirable principle.

In the cases which I have just considered the analytical solution of the question has changed the probability of the causes into certainty. But most often this solution is impossible and it remains only to augment more and more this probability. In the midst of numerous and incalculable modifications which the action of the causes receives then from strange circumstances these causes conserve always with the effects observed the proper ratios to make them recognizable and to verify their existence. Determining these ratios and comparing them with a great number of observations if one finds that they constantly satisfy it, the probability of the causes may increase to the point of equalling that of facts in regard to which there is no doubt. The investigation of these ratios of causes to their effects is not less useful in natural philosophy than the direct solution of problems whether it be to verify the reality of these causes or to determine the laws from their effects; since it may be employed in a great number of questions whose direct solution is not possible, it replaces it in the most advantageous manner. I shall discuss here the application which I have made of it to one of the most interesting phenomena of nature, the flow and the ebb of the sea.

Pliny has given of this phenomenon a description remarkable for its exactitude, and in it one sees that the ancients had observed that the tides of each month are greatest toward the syzygies and smallest toward the quadratures; that they are higher in the perigees than in the apogees of the moon, and higher in the equinoxes than in the solstices. They concluded from this that this phenomenon is due to the action of the sun and moon upon the sea. In the preface of his work *De Stella Martis* Kepler admits a tendency of the waters of the sea toward the moon; but, ignorant of the law of this tendency, he was able to give on this subject only a probable idea. Newton converted into certainty the probability of this idea by attaching it to his great principle of universal gravity. He gave the exact expression of the attractive forces which produced the flood and the ebb of the sea; and in order to determine the effects he supposed that the sea takes at each instant the position of equilibrium which is agreeable to these forces. He explained in this manner the principal phenomena of the tides; but it followed from this theory that in our ports the two tides of the same day would be very unequal if the sun and the moon should have a great declination. At Brest, for example, the evening tide would be in the syzygies of the solstices about eight times greater than the morning tide, which is certainly contrary to the observations which prove that these two tides are very nearly equal. This result from the Newtonian theory might hold to the supposition that the sea is agreeable at each instant to a position of equilibrium, a supposition which is not at all admissible. But the investigation of the true figure of the sea presents great difficulties. Aided by the discoveries which the geometricians had just made in the theory of the movement of fluids and in the calculus of partial differences, I undertook this investigation, and I gave the differential equations of the movement of the sea by supposing that it covers the entire earth. In drawing thus near to nature I had the satisfaction of seeing that my results approached the observations, especially in regard to the little difference which exists in our ports between the two tides of the solstitial syzygies of the same day. I found that they would be equal if the sea had everywhere the same depth; I found further that in giving to this depth convenient values one was able to augment the height of the tides in a port conformably to the observations. But these investigations, in spite of their generality, did not satisfy at all the great differences which even adjacent ports present in this regard and which prove the influence of local circumstances. The impossibility of knowing these circumstances and the irregularity of the basin of the seas and that of integrating the equations of partial differences which are relative has compelled me to make up the deficiency by the method I have indicated above. I then endeavored to determine the greatest ratios possible among the forces which affect all the molecules of the sea; and their effects observable in our ports. For this I made use of the following principle, which may be applied to many other phenomena.

"The state of the system of a body in which the primitive conditions of the movement have disappeared by the resistances which this movement meets is periodic as the forces which animate it."

Combining this principle with that of the coexistence of very small oscillations, I have found an expression of the height of the tides whose arbitraries contain the effect of local cricumstances of each port and are reduced to the smallest number possible; it is only necessary to compare it to a great number of observations.

Upon the invitation of the Academy of Sciences, observations were made at the beginning of the last century at Brest upon the tides, which were continued during six consecutive years. The situation of this port is very favorable to this sort of observations; it communicates with the sea by a canal which empties into a vast roadstead at the far end of which the port has been constructed. The irregularities of the sea extend thus only to a small degree into the port, just as the oscillations which the irregular movement of a vessel produces in a barometer are diminished by a throttling made in the tube of this instrument. Moreover, the tides being considerable at Brest, the accidental variations caused by the winds are only feeble; likewise we notice in the observations of these tides, however little we multiply them, a great regularity which induced me to˙propose to the government to order in this port a new series of observations of the tides, continued during a period of the movement of the nodes of the lunar orbit. This has been done. The observations began June 1, 1806; and since this time they have been made every day without interruption. I am indebted to the indefatigable zeal of M. Bouvard, for all that interests astronomy, the immense calculations which the comparison of my analysis with the observations has demanded. There have been used about six thousand observations, made during the year 1807 and the fifteen years following. It results from this comparison that my formulæ represent with a remarkable precision all the varieties of the tides relative to the digression of the moon, from the sun, to the declination of these stars, to their distances from the earth, and to the laws of variation at the *maximum* and *minimum* of each of these elements. There results from this accord a probability that the flow and the ebb of the sea is due to the attraction of the sun and moon, so approaching certainty that it ought to leave room for no reasonable doubt. It changes into certainty when we consider that this attraction is derived from the law of universal gravity demonstrated by all the celestial phenomena.

The action of the moon upon the sea is more than double that of the sun. Newton and his successors in the development of this action have paid attention only to the terms divided by the cube of the distance from the moon to the earth, judging that the effects due to the following terms ought to be inappreciable. But the calculation of probabilities makes it clear to us that the smallest effects of regular causes may manifest themselves in the results of a great number of observations arranged in the order most suitable to indicate them. This calculation again determines their probability and up to what point it is necessary to multiply the observations to make it very great. Applying it to the numerous observations discussed by M. Bouvard I recognized that at Brest the action of the moon upon the sea is greater in the full moons than in the new moons, and greater when the moon is austral than when it is boreal—phenomena which can result only from the terms of the lunar action divided by the fourth power of the distance from the moon to the earth.

To arrive at the ocean the action of the sun and the moon traverses the atmosphere, which ought consequently to feel its influence and to be subjected to movements similar to those of the sea.

These movements produce in the barometer periodic

oscillations. Analysis has made it clear to me that they are inappreciable in our climates. But as local circumstances increase considerably the tides in our ports, I have inquired again if similar circumstances have made appreciable these oscillations of the barometer. For this I have made use of the meteorological observations which have been made every day for many years at the royal observatory. The heights of the barometer and of the thermometer are observed there at nine o'clock in the morning, at noon, at three o'clock in the afternoon, and at eleven o'clock in the evening. M. Bouvard has indeed wished to take up the consideration of observations of the eight years elapsed from October 1, 1815, to October 1, 1823, on the registers. In disposing the observations in the manner most suitable to indicate the lunar atmospheric flood at Paris, I find only one eighteenth of a millimeter for the extent of the corresponding oscillation of the barometer. It is this especially which has made us feel the necessity of a method for determining the probability of a result, and without this method one is forced to present as the laws of nature the results of irregular causes which has often happened in meteorology. This method applied to the preceding result shows the uncertainty of it in spite of the great number of observations employed, which it would be necessary to increase tenfold in order to obtain a result sufficiently probable.

The principle which serves as a basis for my theory of the tides may be extended to all the effects of hazard to which variable causes are joined according to regular laws. The action of these causes produces in the mean results of a great number of effects varieties which follow the same laws and which one may recognize by the analysis of probabilities. In the measure which these effects are multiplied those varieties are manifested with an ever-increasing probability, which would approach certainty if the number of the effects of the results should become infinite. This theorem is analogous to that which I have already developed upon the action of constant causes. Every time, then, that a cause whose progress is regular can have influence upon a kind of events, we may seek to discover its influence by multiplying the observations and arranging them in the most suitable order to indicate it. When this influence appears to manifest itself the analysis of probabilities determines the probability of its existence and that of its intensity; thus the variation of the temperature from day to night modifying the pressure of the atmosphere and consequently the height of the barometer, it is natural to think that the multiplied observations of these heights ought to show the influence of the solar heat. Indeed there has long been recognized at the equator, where this influence appears to be greatest, a small diurnal variation in the height of the barometer of which the *maximum* occurs about nine o'clock in the morning and the *minimum* about three o'clock in the afternoon. A second *maximum* occurs about eleven o'clock in the evening and a second *minimum* about four o'clock in the morning. The oscillations of the night are less than those of the day, the extent of which is about two millimeters. The inconstancy of our climate has not taken this variation from our observers, although it may be less appreciable than in the tropics. M. Ramond has recognized and determined it at Clermont, the chief place of the district of Puy-de-Dôme, by a series of precise observations made during several years; he has even found that it is smaller in the months of winter than in other months. The numerous observations which I have discussed in order to estimate the influence of attractions of the sun and the moon upon the barometric heights at Paris have served me in determining their diurnal variation. Comparing the heights at nine o'clock in the morning with those of the same days at three o'clock in the afternoon, this variation is manifested with so much evidence that its mean value each month has been constantly positive for each of the seventy-two months from January 1, 1817, to January 1, 1823; its mean value in these seventy-two months has been almost .8 of a millimeter, a little less than at Clermont and much less than at the equator. I have recognized that the mean result of the diurnal variations of the barometer from 9 o'clock A.M. to 3 P.M. has been only .5428 millimeter in the three months of November, December, January, and that it has risen to 1.0563 millimeters in the three following months, which coincides with the observations of M. Ramond. The other months offer nothing similar.

In order to apply to these phenomena the calculation of these probabilities, I commenced by determining the law of the probability of the anomalies of the diurnal variation due to hazard. Applying it then to the observations of this phenomenon, I found that it was a bet of more than 300,000 against one that a regular cause produced it. I do not seek to determine this cause; I content myself with stating its existence. The period of the diurnal variation regulated by the solar day indicates evidently that this variation is due to the action of the sun. The extreme smallness of the attractive action of the sun upon the atmosphere is proved by the smallness of the effects due to the united attractions of the sun and the moon. It is then by the action of its heat that the sun produces the diurnal variation of the barometer; but it is impossible to subject to calculus the effects of its action on the height of the barometer and upon the winds. The diurnal variation of the magnetic needle is certainly a result of the action of the sun. But does this star act here as in the diurnal variation of the barometer by its heat or by its influence upon electricity and upon magnetism, or finally by the union of these influences ? A long series of observations made in different countries will enable us to apprehend this.

One of the most remarkable phenomena of the system of the world is that of all the movemens of rotation and of revolution of the planets and the satellites in the sense of the rotation of the sun and about in the same plane of its equator. A phenomenon so remarkable is not the effect of hazard: it indicates a general cause which has determined all its movements. In order to obtain the probability with which this cause is indicated we shall observe that the planetary system, such as we know it to-day, is composed of eleven planets and of eighteen satellites at least, if we attribute with Herschel six satellites to the planet Uranus. The movements of the rotation of the sun, of six planets, of the moon, of the satellites of

Jupiter, of the ring of Saturn, and of one of its satellites have been recognized. These movements form with those of revolution a totality of forty-three movements directed in the same sense; but one finds by the analysis of probabilities that it is a bet of more than 4000000000000 against one that this disposition is not the result of hazard; this forms a probability indeed superior to that of historical events in regard to which no doubt exists. We ought then to believe at least with equal confidence that a primitive cause has directed the planetary movements, especially if we consider that the inclination of the greatest number of these movements at the solar equator is very small.

Another equally remarkable phenomenon of the solar system is the small degree of the eccentricity of the orbs of the planets and the satellites, while those of the comets are very elongated, the orbs of the system not offering any intermediate shades between a great and a small eccentricity. We are again forced to recognize here the effect of a regular cause; chance has certainly not given an almost circular form to the orbits of all the planets and their satellites; it is then that the cause which has determined the movements of these bodies has rendered them almost circular. It is necessary, again, that the great eccentricities of the orbits of the comets should result from the existence of this cause without its having influenced the direction of their movements; for it is found that there are almost as many retrograde comets as direct comets, and that the mean inclination of all their orbits to the ecliptic approaches very nearly half a right angle, as it ought to be if the bodies had been thrown at hazard.

Whatever may be the nature of the cause in question, since it has produced or directed the movement of the planets, it is necessary that it should have embraced all the bodies and considered all the distances which separate them, it can have been only a fluid of an immense extension. Therefore in order to have given them in the same sense an almost circular movement about the sun it is necessary that this fluid should have surrounded this star as an atmosphere. The consideration of the planetary movements leads us then to think that by virtue of an excessive heat the atmosphere of the sun was originally extended beyond the orbits of all the planets, and that it has contracted gradually to its present limits.

In the primitive state where we imagine the sun it resembled the nebulæ that the telescope shows us composed of a nucleus more or less brilliant surrounded by a nebula which, condensing at the surface, ought to transform it some day into a star. If one conceives by analogy all the stars formed in this manner, one can imagine their anterior state of nebulosity itself preceded by other stars in which the nebulous matter was more and more diffuse, the nucleus being less and less luminous and dense. Going back, then, as far as possible, one would arrive at a nebulosity so diffuse that one would be able scarcely to suspect its existence.

Such is indeed the first state of the nebulæ which Herschel observed with particular care by means of his powerful telescopes, and in which he has followed the progress of condensation, not in a single one, these stages not becoming appreciable to us except after centuries, but in their totality, just about as one can in a vast forest follow the increase of the trees by the individuals of the divers ages which the forest contains. He has observed from the beginning nebulous matter spread out in divers masses in the different parts of the heavens, of which it occupies a great extent. He has seen in some of these masses this matter slightly condensed about one or several faintly luminous nebulæ. In the other nebulæ these nuclei shine, moreover, in proportion to the nebulosity which surrounds them. The atmospheres of each nucleus becoming separated by an ulterior condensation, there result the multifold nebulæ formed of brilliant nuclei very adjacent and surrounded each by an atmosphere; sometimes the nebulous matter, by condensing in a uniform manner, has produced the nebulæ which are called *planetary*. Finally a greater degree of condensation transforms all these nebulæ into stars. The nebulæ classed according to this philosophic view indicate with an extreme probability their future transformation into stars and the anterior state of nebulosity of existing stars. The following considerations come to the aid of proofs drawn from these analogies.

For a long time the particular disposition of certain stars visible to the naked eye has struck the attention of philosophical observers. Mitchel has already remarked how improbable it is that the stars of the Pleiades, for example, should have been confined in the narrow space which contain them by the chances of hazard alone, and he has concluded from this that this group of stars and the similar groups that the heaven presents us are the results of a primitive cause or of a general law of nature. These groups are a necessary result of the condensation of the nebulæ at several nuclei; it is apparent that the nebulous matter being attracted continuously by the divers nuclei, they ought to form in time a group of stars equal to that of the Pleiades. The condensation of the nebulæ at two nuclei forms similarly very adjacent stars, revolving the one about the other, equal to those whose respective movements Herschel has already considered. Such are, further, the 61st of the Swan and its following one in which Bessel has just recognized particular movements so considerable and so little different that the proximity of these stars to one another and their movement about the common centre of gravity ought to leave no doubt. Thus one descends by degrees from the condensation of nebulous matter to the consideration of the sun surrounded formerly by a vast atmosphere, a consideration to which one repasses, as has been seen, by the examination of the phenomena of the solar system. A case so remarkable gives to the existence of this anterior state of the sun a probability strongly approaching certainty.

But how has the solar atmosphere determined the movements of rotation and revolution of the planets and the satellites? If these bodies had penetrated deeply the atmosphere its resistance would have caused them to fall upon the sun; one is then led to believe with much probability that the planets have been formed at the successive limits of the solar atmosphere which, contracting by the cold, ought to have abandoned in the plane of its equator zones of vapors which the mutual attraction of their molecules has changed into divers spheroids. The satellites have been similarly formed by the atmospheres of their respective planets.

I have developed at length in my *Exposition of the System of the World* this hypothesis, which appears to me to satisfy all the phenomena which this system presents us. I shall content myself here with considering that the angular velocity of rotation of the sun and the planets being accelerated by the successive condensation of their atmospheres at their surfaces, it ought to surpass the angular velocity of revolution of the nearest bodies which revolve about them. Observation has indeed confirmed this with regard to the planets and satellites, and even in ratio to the ring of Saturn, the duration of whose revolution is .438 days, while the duration of the rotation of Saturn is .427 days.

In this hypothesis the comets are strangers to the planetary system. In attaching their formation to that of the nebulæ they may be regarded as small nebulæ at the nuclei, wandering from systems to solar systems, and formed by the condensation of the nebulous matter spread out in such great profusion in the universe. The comets would be thus, in relation to our system, as the aerolites are relatively to the Earth, to which they would appear strangers. When these stars become visible to us they offer so perfect resemblance to the nebulæ that they are often confounded with them; and it is only by their movement, or by the knowledge of all the nebulæ confined to that part of the heavens where they appear, that we succeed in distinguishing them. This supposition explains in a happy manner the great extension which the heads and tails of comets take in the measure that they approach the sun, and the extreme rarity of these tails which, in spite of their immense depth, do not weaken at all appreciably the light of the stars which we look across.

When the little nebulæ come into that part of space where the attraction of the sun is predominant, and which we shall call the *sphere of activity* of this star, it forces them to describe elliptic or hyperbolic orbits. But their speed being equally possible in all directions they ought to move indifferently in all the senses and under all inclinations of the elliptic, which is conformable to that which has been observed.

The great eccentricity of the cometary orbits results again from the preceding hypothesis. Indeed if these orbits are elliptical they are very elongated, since their great axes are at least equal to the radius of the sphere of activity of the sun. But these orbits may be hyperbolic; and if the axes of these hyperbolæ are not very large in proportion to the mean distance from the sun to the earth, the movement of the comets which describe them will appear sensibly hyperbolic. However, of the hundred comets of which we already have the elements, not one has appeared certainly to move in an hyperbola; it is necessary, then, that the chances which give an appreciable hyperbola should be extremely rare in proportion to the contrary chances.

The comets are so small that, in order to become visible, their perihelion distance ought to be inconsiderable. Up to the present this distance has surpassed only twice the diameter of the terrestrial orbit, and most often it has been below the radius of this orbit. It is conceived that, in order to approach so near the sun, their speed at the moment of their entrance into its sphere of activity ought to have a magnitude and a direction confined within narrow limits. In determining by the analysis of probabilities the ratio of the chances which, in these limits, give an appreciable hyperbola, to the chances which give an orbit which may be confounded with a parabola, I have found that it is a bet of at least 6000 against one that a nebula which penetrates into the activity of the sun in such a manner as to be observed will describe either a very elongated ellipse or an hyperbola. By the magnitude of its axis, the latter will be appreciably confounded with a parabola in the part which is observed; it is then not surprising that, up to this time, hyperbolic movements have not been recognized.

The attraction of the planets, and, perhaps further, the resistance of the ethereal centres, ought to have changed many cometary orbits in the ellipses whose great axis is less than the radius of the sphere of activity of the sun, which augments the chances of the elliptical orbits. We may believe that this change has taken place with the comet of 1759, and with the comet whose duration is only twelve hundred days, and which will reappear without ceasing in this short interval, unless the evaporation which it meets at each of its returns to the perihelion ends by rendering it invisible.

We are able further, by the analysis of probabilities, to verify the existence or the influence of certain causes whose action is believed to exist upon organized beings. Of all the instruments that we are able to employ in order to recognize the imperceptible agents of nature the most sensitive are the nerves, especially when particular causes increase their sensibility. It is by their aid that the feeble electricity which the contact of two heterogeneous metals develops has been discovered; this has opened a vast field to the researches of physicists and chemists. The singular phenomena which results from extreme sensibility of the nerves in some individuals have given birth to divers opinions about the existence of a new agent which has been named *animal magnetism*, about the action on ordinary magnetism, and about the influence of the sun and moon in some nervous affections, and finally, about the impressions which the proximity of metals or of running water makes felt. It is natural to think that the action of these causes is very feeble, and that it may be easily disturbed by accidental circumstances; thus because in some cases it is not manifested at all its existence ought not to be denied. We are so far from recognizing all the agents of nature and their divers modes of action that it would be unphilosophical to deny the phenomena solely because they are inexplicable in the present state of our knowledge. But we ought to examine them with an attention as much the more scrupulous as it appears the more difficult to admit them; and it is here that the calculation of probabilities becomes indispensable in determining to just what point it is necessary to multiply the observations or the experiences in order to obtain in favor of the agents which they indicate, a probability superior to the reasons which can be obtained elsewhere for not admitting them.

The calculation of probabilities can make appreciable the advantages and the inconveniences of the methods employed in the speculative sciences. Thus in order to recognize the best of the treatments in use in the healing of a malady, it is sufficient to test each of them on an equal number of patients, making all the conditions exactly similar; the superiority of the most advantageous treatment will manifest itself more and more in the measure that the number is increased; and the calculation will make apparent the corresponding probability of its advantage and the ratio according to which it is superior to the others.

CHAPTER X.

APPLICATION OF THE CALCULUS OF PROB-ABILITIES TO THE MORAL SCIENCES.

WE have just seen the advantages of the analysis of probabilities in the investigation of the laws of natural phenomena whose causes are unknown or so complicated that their results cannot be submitted to calculus. This is the case of nearly all subjects of the moral sciences. So many unforeseen causes, either hidden or inappreciable, influence human institutions that it is impossible to judge *à priori* the results. The series of events which time brings about develops these results and indicates the means of remedying those that are harmful. Wise laws have often been made in this regard; but because we had neglected to conserve the motives many have been abrogated as useless, and the fact that vexatious experiences have made the need felt anew ought to have reëstablished them.

It is very important to keep in each branch of the public administration an exact register of the results which the various means used have produced, and which are so many experiences made on a large scale by governments. Let us apply to the political and moral sciences the method founded upon observation and upon calculus, the method which has served us so well in the natural sciences. Let us not offer in the least a useless and often dangerous resistance to the inevitable effects of the progress of knowledge; but let us change only with an extreme circumspection our institutions and the usages to which we have already so long conformed. We should know well by the experience of the past the difficulties which they present; but we are ignorant of the extent of the evils which their change can produce. In this ignorance the theory of probability directs us to avoid all change; especially is it necessary to avoid the sudden changes which in the moral world as well as in the physical world never operate without a great loss of vital force.

Already the calculus of probabilities has been applied with success to several subjects of the moral sciences. I shall present here the principal results.

CHAPTER XI.

CONCERNING THE PROBABILITIES OF TESTI-MONIES.

THE majority of our opinions being founded on the probability of proofs it is indeed important to submit it to calculus. Things it is true often become impossible by the difficulty of appreciating the veracity of witnesses and by the great number of circumstances which accompany the deeds they attest; but one is able in several cases to resolve the problems which have much analogy with the questions which are proposed and whose solutions may be regarded as suitable approximations to guide and to defend us againt the errors and the dangers of false reasoning to which we are exposed. An approximation of this kind, when it is well made, is always preferable to the most specious reasonings. Let us try then to give some general rules for obtaining it.

A single number has been drawn from an urn which contains a thousand of them. A witness to this drawing announces that number 79 is drawn; one asks the probability of drawing this number. Let us suppose that experience has made known that this witness deceives one time in ten, so that the probability of his testimony is $\frac{1}{10}$. Here the event observed is the witness attesting that number 79 is drawn. This event may result from the two following hypotheses, namely: that the witness utters the truth or that he deceives. Following the principle that has been expounded on the probability of causes drawn from events observed it is necessary first to determine *à priori* the probability of the event in each hypothesis. In the first, the probability that the witness will announce number 79 is the probability itself of the drawing of this number, that is to say, $\frac{1}{1000}$. It is necessary to multiply it by the probability $\frac{9}{10}$ of the veracity of the witness; one will have then $\frac{9}{1000}$ for the probability of the event observed in this hypothesis. If the witness deceives, number 79 is not drawn, and the probability of this case is $\frac{999}{1000}$. But to announce the drawing of this number the witness has to choose it among the 999 numbers not drawn; and as he is supposed to have no motive of preference for the ones rather than the others, the probability that he will choose number 79 is $\frac{1}{999}$; multiplying, then, this probability by the preceding one, we shall have $\frac{1}{1000}$ for the probability that the witness will announce number 79 in the second hypothesis. It is necessary again to multiply this probability by $\frac{1}{10}$ of the hypothesis itself, which gives $\frac{1}{10000}$ for the probability of the event relative to this hypothesis. Now if we form a fraction whose numerator is the probability relative to the first hypothesis, and whose denominator is the sum of the probabilities relative to the two hypotheses, we shall have, by the sixth principle, the probability of the first hypothesis, and this probability will be $\frac{9}{10}$; that is to say, the veracity itself of the witness. This is likewise the probability of the drawing of number 79. The probability of the falsehood of the witness and of the failure of drawing this number is $\frac{1}{10}$.

If the witness, wishing to deceive, has some interest in choosing number 79 among the numbers not drawn, —if he judges, for example, that having placed upon this number a considerable stake, the announcement of its drawing will increase his credit, the probability that he will choose this number will no longer be as at first, $\frac{1}{999}$, it will then be $\frac{1}{2}$, $\frac{1}{3}$, etc., according to the interest that he will have in announcing its drawing. Supposing it to be $\frac{1}{9}$, it will be necessary to multiply by this fraction the probability $\frac{999}{1000}$ in order to get in the hypothesis of the falsehood the probability of the event observed, which it is necessary still to multiply by $\frac{1}{10}$, which gives $\frac{111}{10000}$ for the probability of the event in the second hypothesis. Then the probability of the first hypothesis, or of the drawing of number 79, is reduced by the preceding rule to $\frac{9}{120}$. It is then very much decreased by the consideration of the interest which the witness may have in announcing the drawing of number 79. In truth this same interest increases the probability $\frac{9}{10}$ that the witness will speak the truth if number 79 is drawn. But this probability cannot exceed unity or $\frac{10}{10}$; thus the probability of the drawing of number 79 will not surpass $\frac{10}{121}$. Common sense tells us that this interest ought to inspire distrust, but calculus appreciates the influence of it.

The probability *à priori* of the number announced by the witness is unity divided by the number of the numbers in the urn; it is changed by virtue of the proof into the veracity itself of the witness; it may then be decreased by the proof. If, for example, the urn contains only two numbers, which gives $\frac{1}{2}$ for the probability *à priori* of the drawing of number 1, and if the veracity of a witness who announces it is $\frac{4}{10}$, this drawing becomes less probable. Indeed it is apparent, since the witness has then more inclination towards a falsehood than towards the truth, that his testimony ought to decrease the probability of the fact attested every time that this probability equals or surpasses $\frac{1}{2}$. But if there are three numbers in the urn the probability *à priori* of the drawing of number 1 is increased by the affirmation of a witness whose veracity surpasses $\frac{1}{3}$.

Suppose now that the urn contains 999 black balls and one white ball, and that one ball having been drawn a witness of the drawing announces that this ball is white. The probability of the event observed, determined *à priori* in the first hypothesis, will be here, as in the preceding question, equal to $\frac{9}{10000}$. But in the hypothesis where the witness deceives, the white ball is not drawn and the probability of this case is $\frac{999}{1000}$. It is necessary to multiply it by the probability $\frac{1}{10}$ of the falsehood, which gives $\frac{999}{10000}$ for the probability of the event observed relative to the second hypothesis. This probability was only $\frac{1}{10000}$ in the preceding question; this great difference results from this—that a black ball having been drawn the witness who wishes to deceive has no choice at all to make among the 999 balls not drawn in order to announce the drawing of a white ball. Now if one forms two fractions whose numerators are the probabilities relative to each hypothesis, and whose common denominator is the sum of these probabilities, one will have $\frac{9}{1008}$ for the probability of the first hypothesis and of the drawing of a white ball, and $\frac{999}{1008}$ for the probability of the second hypothesis and of the drawing of a black ball. This last probability strongly approaches certainty; it would approach it much nearer and would become $\frac{999999}{1000008}$ if the urn contained a million balls of which one was white, the drawing of a white ball becoming then much more extraordinary. We see thus how the probability of the falsehood increases in the measure that the deed becomes more extraordinary.

We have supposed up to this time that the witness was not mistaken at all; but if one admits, however, the chance of his error the extraordinary incident becomes more improbable. Then in place of the two hypotheses one will have the four following ones, namely: that of the witness not deceiving and not being mistaken at all; that of the witness not deceiving at all and being mistaken; the hypothesis of the witness deceiving and not being mistaken at all; finally, that of the witness deceiving and being mistaken. Determining *à priori* in each of these hypotheses the probability of the event observed, we find by the sixth principle the probability that the fact attested is false equal to a fraction whose numerator is the number of black balls in the urn multiplied by the sum of the probabilities that the witness does not deceive at all and is mistaken, or that he deceives and is not mistaken, and whose denominator is this numerator augmented by the sum of the probabilities that the witness does not deceive at all and is not mistaken at all, or that he deceives and is mistaken at the same time. We see by this that if the number of black balls in the urn is very great, which renders the drawing of the white ball extraordinary, the probability that the fact attested is not true approaches most nearly to certainty.

Applying this conclusion to all extraordinary deeds it results from it that the probability of the error or of the falsehood of the witness becomes as much greater as the fact attested is more extraordinary. Some authors have advanced the contrary on this basis that the view of an extraordinary fact being perfectly similar to that of an ordinary fact the same motives ought to lead us to give the witness the same credence when he affirms the one or the other of these facts. Simple common sense rejects such a strange assertion; but the calculus of probabilities, while confirming the findings of common sense, appreciates the greatest improbability of testimonies in regard to extraordinary facts.

These authors insist and suppose two witnesses equally worthy of belief, of whom the first attests that he saw an individual dead fifteen days ago whom the second witness affirms to have seen yesterday full of life. The one or the other of these facts offers no improbability. The reservation of the individual is a result of their combination; but the testimonies do not bring us at all directly to this result, although the credence which is due these testimonies ought not to be decreased by the fact that the result of their combination is extraordinary.

But if the conclusion which results from the combination of the testimonies was impossible one of them would be necessarily false; but an impossible conclusion is the limit of extraordinary conclusions, as error is the limit of improbable conclusions; the value of the testimonies which becomes zero in the case of an impossible conclusion ought then to be very much decreased in that of an extraordinary conclusion. This is indeed confirmed by the calculus of probabilities.

In order to make it plain let us consider two urns, A and B, of which the first contains a million white balls and the second a million black balls. One draws from one of these urns a ball, which he puts back into the other urn, from which one then draws a ball. Two witnesses, the one of the first drawing, the other of the second, attest that the ball which they have seen drawn is white without indicating the urn from which it has been drawn. Each testimony taken alone is not improbable; and it is easy to see that the probability of the fact attested is the veracity itself of the witness. But it follows from the combination of the testimonies that a white ball has been extracted from the urn A at the first draw, and that then placed in the urn B it has reappeared at the second draw, which is very extraordinary; for this second urn, containing then one white ball among a million black balls, the probability of drawing the white ball is $\frac{1}{1000001}$. In order to determine the diminution which results in the probability of the thing announced by the two witnesses we shall notice that the event observed is here the affirmation by each of them that the ball which he has seen extracted is white. Let us represent by $\frac{9}{10}$ the probability that he announces the truth, which can occur in the present case when the witness does not deceive and is not mistaken at all, and when he deceives and is mistaken at the same time. One may form the four following hypotheses:

1st. The first and second witness speak the truth. Then a white ball has at first been drawn from the urn A, and the probability of this event is $\frac{1}{2}$, since the ball drawn at the first draw may have been drawn either from the one or the other urn. Consequently the ball drawn, placed in the urn B, has reappeared at the second draw; the probability of this event is $\frac{1}{1000001}$, the probability of the fact announced is then $\frac{1}{2000002}$. Multiplying it by the product of the probabilities $\frac{9}{10}$ and $\frac{9}{10}$ that the witnesses speak the truth one will have $\frac{81}{200000020200}$ for the probability of the event observed in this first hypothesis.

2d. The first witness speaks the truth and the second does not, whether he deceives and is not mistaken or he does not deceive and is mistaken. Then a white ball has been drawn from the urn A at the first draw, and the probability of this event is $\frac{1}{2}$. Then this ball having been placed in the urn B a black ball has been drawn from it: the probability of such drawing is $\frac{1000000}{1000001}$; one has then $\frac{1000000}{2000002}$ for the probability of the compound event. Multiplying it by the product of the two probabilities $\frac{9}{10}$ and $\frac{1}{10}$ that the first witness speaks the truth and that the second does not, one will have $\frac{9000000}{200000020200}$ for the probability for the event observed in the second hypothesis.

3d. The first witness does not speak the truth and the second announces it. Then a black ball has been drawn from the urn B at the first drawing, and after having been placed in the urn A a white ball has been drawn from this urn. The probability of the first of these events is $\frac{1}{2}$ and that of the second is $\frac{1000000}{10000001}$; the probability of the compound event is then $\frac{1000000}{20000002}$. Multiplying it by the product of the probabilities $\frac{1}{10}$ and $\frac{9}{10}$ that the first witness does not speak the truth and that the second announces it, one will have $\frac{9000000}{2000000200}$ for the probability of the event observed relative to this hypothesis.

4th. Finally, neither of the witnesses speaks the truth. Then a black ball has been drawn from the urn B at the first draw; then having been placed in the urn A it has reappeared at the second drawing: the probability of this compound event is $\frac{1}{20000002}$. Multiplying it by the product of the probabilities $\frac{1}{10}$ and $\frac{1}{10}$ that each witness does not speak the truth one will have $\frac{1}{2000000200}$ for the probability of the event observed in this hypothesis.

Now in order to obtain the probability of the thing announced by the two witnesses, namely, that a white ball has been drawn at each draw, it is necessary to divide the probability corresponding to the first hypothesis by the sum of the probabilities relative to the four hypotheses; and then one has for this probability $\frac{81}{1800000082}$, an extremely small fraction.

If the two witnesses affirm the first, that a white ball has been drawn from one of the two urns A and B; the second that a white ball has been likewise drawn from one of the two urns A' and B', quite similar to the first ones, the probability of the thing announced by the two witnesses will be the product of the probabilities of their testimonies, or $\frac{81}{100}$; it will then be at least a hundred and eighty thousand times greater than the preceding one. One sees by this how much, in the first case, the reappearance at the second draw of the white ball drawn at the first draw, the extraordinary conclusion of the two testimonies decreases the value of it.

We would give no credence to the testimony of a man who should attest to us that in throwing a hundred dice into the air they had all fallen on the same face. If we had ourselves been spectators of this event we should believe our own eyes only after having carefully examined all the circumstances, and after having brought in the testimonies of other eyes in order to be quite sure that there had been neither hallucination nor deception. But after this examination we should not hesitate to admit it in spite of its extreme improbability; and no one would be tempted, in order to explain it, to recur to a denial of the laws of vision. We ought to conclude from it that the probability of the constancy of the laws of nature is for us greater than this, that the event in question has not taken place at all—a probability greater than that of the majority of historical facts which we regard as incontestable. One may judge by this the immense weight of testimonies necessary to admit a suspension of natural laws, and how improper it would be to apply to this case the ordinary rules of criticism. All those who without offering this immensity of testimonies support this when making recitals of events contrary to those laws, decrease rather than augment the belief which they wish to inspire; for then those recitals render very probable the error or the falsehood of their authors.

But that which diminishes the belief of educated men increases often that of the uneducated, always greedy for the wonderful.

There are things so extraordinary that nothing can balance their improbability. But this, by the effect of a dominant opinion, can be weakened to the point of appearing inferior to the probability of the testimonies; and when this opinion changes an absurd statement admitted unanimously in the century which has given it birth offers to the following centuries only a new proof of the extreme influence of the general opinion upon the more enlightened minds. Two great men of the century of Louis XIV.—Racine and Pascal—are striking examples of this. It is painful to see with what complaisance Racine, this admirable painter of the human heart and the most perfect poet that has ever lived, reports as miraculous the recovery of Mlle. Perrier, a niece of Pascal and a day pupil at the monastery of Port-Royal; it is painful to read the reasons by which Pascal seeks to prove that this miracle should be necessary to religion in order to justify the doctrine of the monks of this abbey, at that time persecuted by the Jesuits. The young Perrier had been afflicted for three years and a half by a lachrymal fistula; she touched her afflicted eye with a relic which was pretended to be one of the thorns of the crown of the Saviour and she had faith in instant recovery. Some days afterward the physicians and the surgeons attest the recovery, and they declare that nature and the remedies have had no part in it. This event, which took place in 1656, made a great sensation, and "all Paris rushed," says Racine, "to Port-Royal. The crowd increased from day to day, and God himself seemed to take pleasure in authorizing the devotion of the people by the number of miracles which were performed in this church.'' At this time miracles and sorcery did not yet appear improbable, and one did not hesitate at all to attribute to them the singularities of nature which could not be explained otherwise.

This manner of viewing extraordinary results is found in the most remarkable works of the century of Louis XIV.; even in the Essay on the Human Understanding by the philosopher Locke, who says, in speaking of the degree of assent: '' Though the common experience and the ordinary course of things have justly a mighty influence on the minds of men, to make them give or refuse credit to anything proposed to their belief; yet there is one case, wherein the strangeness of the fact lessens not the assent to a fair testimony of it. For where such supernatural events are suitable to ends aimed at by him who has the power to change the course of nature, there, under such circumstances, they may be the fitter to procure belief, by how much the more they are beyond or contrary to ordinary observation.'' The true principles of the probability of testimonies having been thus misunderstood by philosophers to whom reason is principally indebted for its progress, I have thought it necessary to present at length the results of calculus upon this important subject.

There comes up naturally at this point the discussion of a famous argument of Pascal, that Craig, an English mathematician, has produced under a geometric form. Witnesses declare that they have it from Divinity that in conforming to a certain thing one will enjoy not one or two but an infinity of happy lives. However feeble the probability of the proofs may be, provided that it be not infinitely small, it is clear that the advantage of those who conform to the prescribed thing is infinite since it is the product of this probability and an infinite good; one ought not to hesitate then to procure for oneself this advantage.

This argument is based upon the infinite number of happy lives promised in the name of the Divinity by the witnesses; it is necessary then to prescribe them, precisely because they exaggerate their promises beyond all limits, a consequence which is repugnant to good sense. Also calculus teaches us that this exaggeration itself enfeebles the probability of their testimony to the point of rendering it infinitely small or zero. Indeed this case is similar to that of a witness who should announce the drawing of the highest number from an urn filled with a great number of numbers, one of which has been drawn and who would have a great interest in announcing the drawing of this number. One has already seen how much this interest enfeebles his testimony. In evaluating only at $\frac{1}{2}$ the probability that if the witness deceives he will choose the largest number, calculus gives the probability of his announcement as smaller than a fraction whose numerator is unity and whose denominator is unity plus the half of the product of the number of the numbers by the probability of falsehood considered *à priori* or independently of the announcement. In order to compare this case to that of the argument of Pascal it is sufficient to represent by the numbers in the urn all the possible numbers of happy lives which the number

of these numbers renders infinite; and to observe that
if the witnesses deceive they have the greatest interest,
in order to accredit their falsehood, in promising an
eternity of happiness. The expression of the probability of their testimony becomes then infinitely small.
Multiplying it by the infinite number of happy lives
promised, infinity would disappear from the product
which expresses the advantage resultant from this
promise which destroys the argument of Pascal.

Let us consider now the probability of the totality
of several testimonies upon an established fact. In
order to fix our ideas let us suppose that the fact be
the drawing of a number from an urn which contains a
hundred of them, and of which one single number has
been drawn. Two witnesses of this drawing announce
that number 2 has been drawn, and one asks for the
resultant probability of the totality of these testimonies.
One may form these two hypotheses: the witnesses
speak the truth; the witnesses deceive. In the first
hypothesis the number 2 is drawn and the probability
of this event is $\frac{1}{100}$. It is necessary to multiply it by
the product of the veracities of the witnesses, veracities
which we will suppose to be $\frac{9}{10}$ and $\frac{7}{10}$: one will have
then $\frac{63}{10000}$ for the probability of the event observed in
this hypothesis. In the second, the number 2 is not
drawn and the probability of this event is $\frac{99}{100}$. But
the agreement of the witnesses requires then that in
seeking to deceive they both choose the number 2 from
the 99 numbers not drawn: the probability of this
choice if the witnesses do not have a secret agreement
is the product of the fraction $\frac{1}{99}$ by itself; it becomes
necessary then to multiply these two probabilities

together, and by the product of the probabilities $\frac{1}{10}$ and $\frac{3}{10}$ that the witnesses deceive; one will have thus $\frac{1}{330000}$ for the probability of the event observed in the second hypothesis. Now one will have the probability of the fact attested or of the drawing of number 2 in dividing the probability relative to the first hypothesis by the sum of the probabilities relative to the two hypotheses; this probability will be then $\frac{2079}{2080}$, and the probability of the failure to draw this number and of the falsehood of the witnesses will be $\frac{1}{2080}$.

If the urn should contain only the numbers 1 and 2 one would find in the same manner $\frac{21}{22}$ for the probability of the drawing of number 2, and consequently $\frac{1}{22}$ for the probability of the falsehood of the witnesses, a probability at least ninety-four times larger than the preceding one. One sees by this how much the probability of the falsehood of the witnesses diminishes when the fact which they attest is less probable in itself. Indeed one conceives that then the accord of the witnesses, when they deceive, becomes more difficult, at least when they do not have a secret agreement, which we do not suppose here at all.

In the preceding case where the urn contained only two numbers the *à priori* probability of the fact attested is $\frac{1}{2}$, the resultant probability of the testimonies is the product of the veracities of the witnesses divided by this product added to that of the respective probabilities of their falsehood.

It now remains for us to consider the influence of time upon the probability of facts transmitted by a traditional chain of witnesses. It is clear that this probability ought to diminish in proportion as the chain is prolonged. If the fact has no probability itself, such as the drawing of a number from an urn which contains an infinity of them, that which it acquires by the testimonies decreases according to the continued product of the veracity of the witnesses. If the fact has a probability in itself; if, for example, this fact is the drawing of the number 2 from an urn which contains an infinity of them, and of which it is certain that one has drawn a single number; that which the traditional chain adds to this probability decreases, following a continued product of which the first factor is the ratio of the number of numbers in the urn less one to the same number, and of which each other factor is the veracity of each witness diminished by the ratio of the probability of his falsehood to the number of the numbers in the urn less one; so that the limit of the probability of the fact is that of this fact considered *à priori*, or independently of the testimonies, a probability equal to unity divided by the number of the numbers in the urn.

The action of time enfeebles then, without ceasing, the probability of historical facts just as it changes the most durable monuments. One can indeed diminish it by multiplying and conserving the testimonies and the monuments which support them. Printing offers for this purpose a great means, unfortunately unknown to the ancients. In spite of the infinite advantages which it procures the physical and moral revolutions by which the surface of this globe will always be agitated will end, in conjunction with the inevitable effect of time, by rendering doubtful after thousands of years the historical facts regarded to-day as the most certain.

Craig has tried to submit to calculus the gradual enfeebling of the proofs of the Christian religion; supposing that the world ought to end at the epoch when it will cease to be probable, he finds that this ought to take place 1454 years after the time when he writes. But his analysis is as faulty as his hypothesis upon the duration of the moon is bizarre.

CHAPTER XII.

CONCERNING THE SELECTIONS AND THE DECISIONS OF ASSEMBLIES.

THE probability of the decisions of an assembly depends upon the plurality of votes, the intelligence and the impartiality of the members who compose it. So many passions and particular interests so often add their influence that it is impossible to submit this probability to calculus. There are, however, some general results dictated by simple common sense and confirmed by calculus. If, for example, the assembly is poorly informed about the subject submitted to its decision, if this subject requires delicate considerations, or if the truth on this point is contrary to established prejudices, so that it would be a bet of more than one against one that each voter will err; then the decision of the majority will be probably wrong, and the fear of it will be the better based as the assembly is more numerous. It is important then, in public affairs, that assemblies should have to pass upon subjects within reach of the greatest number; it is important for them that information be generally diffused and that good works founded upon reason and experience should enlighten those

who are called to decide the lot of their fellows or to
govern them, and should forewarn them against false
ideas and the prejudices of ignorance. Scholars have
had frequent occasion to remark that first conceptions
often deceive and that the truth is not always probable.

It is difficult to understand and to define the desire
of an assembly in the midst of a variety of opinions of
its members. Let us attempt to give some rules in
regard to this matter by considering the two most
ordinary cases: the election among several candidates,
and that among several propositions relative to the
same subject.

When an assembly has to choose among several
candidates who present themselves for one or for several
places of the same kind, that which appears simplest
is to have each voter write upon a ticket the names of
all the candidates according to the order of merit that
he attributes to them. Supposing that he classifies
them in good faith, the inspection of these tickets will
give the results of the elections in such a manner that
the candidates may be compared among themselves;
so that new elections can give nothing more in this
regard. It is a question now to conclude the order of
preference which the tickets establish among the candidates. Let us imagine that one gives to each voter an
urn which contains an infinity of balls by means of
which he is able to shade all the degrees of merit of
the candidates; let us conceive again that he draws
from his urn a number of balls proportional to the
merit of each candidate, and let us suppose this number
written upon a ticket at the side of the name of the
candidate. It is clear that by making a sum of all the

numbers relative to each candidate upon each ticket, that one of all the candidates who shall have the largest sum will be the candidate whom the assembly prefers; and that in general the order of preference of the candidates will be that of the sums relative to each of them.   But the tickets do not mark at all the number of balls which each voter gives to the candidates; they indicate solely that the first has more of them than the second, the second more than the third, and so on. In supposing then at first upon a given ticket a certain number of balls all the combinations of the inferior numbers which fulfil the preceding conditions are equally admissible; and one will have the number of balls relative to each candidate by making a sum ot all the numbers which each combination gives him and dividing it by the entire number of combinations.   A very simple analysis shows that the numbers which must be written upon each ticket at the side of the last name, of the one before the last, etc., are proportional to the terms of the arithmetical progression 1, 2, 3, etc.   Writing then thus upon each ticket the terms of this progression, and adding the terms relative to each candidate upon these tickets, the divers sums will indicate by their magnitude the order of their preference which ought to be established among the candidates. Such is the mode of election which The Theory of Probabilities indicates.   Without doubt it would be better if each voter should write upon his ticket the names of the candidates in the order of merit which he attributes to them.   But particular interests and many strange considerations of merit would affect this order and place sometimes in the last rank the candidate most formidable to that one whom one prefers, which gives too great an advantage to the candidates of mediocre merit. Likewise experience has caused the abandonment of this mode of election in the societies which had adopted it.

The election by the absolute majority of the suffrages unites to the certainty of not admitting any one of the candidates whom this majority rejects, the advantage of expressing most often the desire of the assembly. It always coincides with the preceding mode when there are only two candidates. Indeed it exposes an assembly to the inconvenience of rendering elections interminable. But experience has shown that this inconvenience is nil, and that the general desire to put an end to elections soon unites the majority of the suffrages upon one of the candidates.

The choice among several propositions relative to the same object ought to be subjected, seemingly, to the same rules as the election among several candidates. But there exists between the two cases this difference, namely, that the merit of a candidate does not exclude that of his competitors; but if it is necessary to choose among propositions which are contrary, the truth of the one excludes the truth of the others. Let us see how one ought then to view this question.

Let us give to each voter an urn which contains an infinite number of balls, and let us suppose that he distributes them upon the divers propositions according to the respective probabilities which he attributes to them. It is clear that the total number of balls expressing certainty, and the voter being by the hypothesis assured that one of the propositions ought to be true, he will distribute this number at length upon the propositions. The problem is reduced then to this, namely, to determine the combinations in which the balls will be distributed in such a manner that there may be more of them upon the first proposition of the ticket than upon the second, more upon the second than upon the third, etc. ; to make the sums of all the numbers of balls relative to each proposition in the divers combinations, and to divide this sum by the number of combinations; the quotients will be the numbers of balls that one ought to attribute to the propositions upon a certain ticket. One finds by analysis that in going from the last proposition these quotients are among themselves as the following quantities : first, unity divided by the number of propositions; second, the preceding quantity, augmented by unity, divided by the number of propositions less one; third, this second quantity, augmented by unity, divided by the number of propositions less two, and so on for the others. One will write then upon each ticket these quantities at the side of the corresponding propositions, and adding the relative quantities to each proposition upon the divers tickets the sums will indicate by their magnitude the order of preference which the assembly gives to these propositions.

Let us speak a word about the manner of renewing assemblies which should change in totality in a definite number of years. Ought the renewal to be made at one time, or is it advantageous to divide it among these years ? According to the last method the assembly would be formed under the influence of the divers opinions dominant during the time of its renewal; the opinion which obtained then would be probably the mean of all these opinions. The assembly would receive thus at the time the same advantage that is given to it by the extension of the elections of its members to all parts of the territory which it represents. Now if one considers what experience has only too clearly taught, namely, that elections are always directed in the greatest degree by dominant opinions, one will feel how useful it is to temper these opinions, the ones by the others, by means of a partial renewal.

CHAPTER XIII.

CONCERNING THE PROBABILITY OF THE JUDG-MENTS OF TRIBUNALS.

ANALYSIS confirms what simple common sense teaches us, namely, the correctness of judgments is as much more probable as the judges are more numerous and more enlightened. It is important then that tribunals of appeal should fulfil these two conditions. The tribunals of the first instance standing in closer relation to those amenable offer to the higher tribunal the advantage of a first judgment already probable, and with which the latter often agree, be it in compromising or in desisting from their claims. But if the uncertainty of the matter in litigation and its importance determine a litigant to have recourse to the tribunal of appeals, he ought to find in a greater probability of obtaining an equitable judgment greater security for his fortune and the compensation for the trouble and expense which a new procedure entails. It is this which had no place in the institution of the reciprocal appeal of the tribunals of the district, an institution thereby very prejudicial to the interest of the citizens. It would be perhaps proper and conformable to the calculus of probabilities to demand a majority of at least two votes in a tribunal of appeal in order to invalidate the sentence of the lower tribunal. One would obtain this result if the tribunal of appeal being composed of an even number of judges the sentence should stand in the case of the equality of votes.

I shall consider particularly the judgments in criminal matters.

In order to condemn an accused it is necessary without doubt that the judges should have the strongest proofs of his offence. But a moral proof is never more than a probability; and experience has only too clearly shown the errors of which criminal judgments, even those which appear to be the most just, are still susceptible. The impossibility of amending these errors is the strongest argument of the philosophers who have wished to proscribe the penalty of death. We should then be obliged to abstain from judging if it were necessary for us to await mathematical evidence. But the judgment is required by the danger which would result from the impunity of the crime. This judgment reduces itself, if I am not mistaken, to the solution of the following question: Has the proof of the offence of the accused the high degree of probability necessary so that the citizens would have less reason to doubt the errors of the tribunals, if he is innocent and condemned, than they would have to fear his new crimes and those of the unfortunate ones who would be emboldened by the example of his impunity if he were guilty and acquitted? The solution of this question depends upon several elements very difficult to ascertain. Such is the eminence of danger which would threaten society if the criminal accused should remain unpunished. Sometimes this danger is so great that the magistrate sees himself constrained to waive forms wisely established for the protection of innocence. But that which renders almost always this question insoluble is the impossibility of appreciating exactly the probability of the offence and of fixing that which is necessary for the condemnation of the accused. Each judge in this respect is forced to rely upon his own judgment. He forms his opinion by comparing the divers testimonies and the circumstances by which the offence is accompanied, to the results of his reflections and his experiences, and in this respect a long habitude of interrogating and judging accused persons gives great advantage in ascertaining the truth in the midst of indices often contradictory.

The preceding question depends again upon the care taken in the investigation of the offence; for one demands naturally much stronger proofs for imposing the death penalty than for inflicting a detention of some months. It is a reason for proportioning the care to the offence, great care taken with an unimportant case inevitably clearing many guilty ones. A law which gives to the judges power of moderating the care in the case of attenuating circumstances is then conformable at the same time to principles of humanity towards the culprit, and to the interest of society. The product of the probability of the offence by its gravity being the measure of the danger to which the acquittal of the accused can expose society, one would think that the care taken ought to depend upon this probability. This is done indirectly in the tribunals where one retains for some time the accused against whom there are very strong proofs, but insufficient to condemn him; in the hope of acquiring new light one does not place him immediately in the midst of his fellow citizens, who would not see him again without great alarm. But the arbitrariness of this measure and the abuse which one can make of it have caused its rejection in the countries where one attaches the greatest price to individual liberty.

Now what is the probability that the decision of a tribunal which can condemn only by a given majority will be just, that is to say, conform to the true solution of the question proposed above? This important problem well solved will give the means of comparing among themselves the different tribunals. The majority of a single vote in a numerous tribunal indicates that the affair in question is very doubtful; the condemnation of the accused would be then contrary to the principles of humanity, protectors of innocence. The unanimity of the judges would give very strong probability of a just decision; but in abstaining from it too many guilty ones would be acquitted. It is necessary, then, either to limit the number of judges, if one wishes that they should be unanimous, or increase the majority necessary for a condemnation, when the tribunal becomes more numerous. I shall attempt to apply calculus to this subject, being persuaded that it is always the best guide when one bases it upon the data which common sense suggests to us.

The probability that the opinion of each judge is just enters as the principal element into this calculation. If in a tribunal of a thousand and one judges, five hundred and one are of one opinion, and five hundred are of the contrary opinion, it is apparent that the probability of the opinion of each judge surpasses very little $\frac{1}{2}$; for supposing it obviously very large a single vote of difference would be an improbable event. But if the judges are unanimous, this indicates in the proofs that degree of strength which entails conviction; the probability of the opinion of each judge is then very near unity or certainty, provided that the passions or the ordinary prejudices do not affect at the same time all the judges. Outside of these cases the ratio of the votes for or against the accused ought alone to determine this probability. I suppose thus that it can vary from $\frac{1}{2}$ to unity, but that it cannot be below $\frac{1}{2}$. If that were not the case the decision of the tribunal would be as insignificant as chance; it has value only in so far as the opinion of the judge has a greater tendency to truth than to error. It is thus by the ratio of the numbers of votes favorable, and contrary to the accused, that I determine the probability of this opinion.

These data suffice to ascertain the general expression of the probability that the decision of a tribunal judging by a known majority is just. In the tribunals where of eight judges five votes would be necessary for the condemnation of an accused, the probability of the error to be feared in the justice of the decision would surpass $\frac{1}{4}$. If the tribunal should be reduced to six members who are able to condemn only by a plurality of four votes, the probability of the error to be feared would be below $\frac{1}{4}$. There would be then for the accused an advantage in this reduction of the tribunal. In both cases the majority required is the same and is equal to two. Thus the majority remaining constant, the probability of error increases with the number of judges; this is general whatever may be the majority required, provided that it remains the same. Taking, then, for the rule the arithmetical ratio, the accused finds himself in a position less and less advantageous in the measure that the tribunal becomes more numerous. One might believe that in a tribunal where one might demand a majority of twelve votes, whatever the number of the judges was, the votes of the minority, neutralizing an equal number of votes of the majority, the twelve remaining votes would represent the unanimity of a jury of twelve members, required in England for the condemnation of an accused; but one would be greatly mistaken. Common sense shows that there is a difference between the decision of a tribunal of two hundred and twelve judges, of which one hundred and twelve condemn the accused, while one hundred acquit him, and that of a tribunal of twelve judges unanimous for condemnation. In the first case the hundred votes favorable to the accused warrant in thinking that the proofs are far from attaining the degree of strength which entails conviction; in the second case, the unanimity of the judges leads to the belief that they have attained this degree. But simple common sense does not suffice at all to appreciate the extreme difference of the probability of error in the two cases. It is necessary then to recur to calculus, and one finds nearly one fifth for the probability of error in the first case, and only $\frac{1}{8192}$ for this probability in the second case, a probability which is not one thousandth of the first. It is a confirmation of the principle that the arithmetical ratio is unfavorable to the accused when the number of judges increases. On the contrary, if one takes for a rule the geometrical ratio, the probability of the error of the decision diminishes when the number of judges increases. For example, in the tribunals which can condemn only by a plurality of two thirds of the votes, the probability of the error to be feared is nearly one fourth if the number of the judges is six; it is below $\frac{1}{7}$ if this number is increased to twelve. Thus one ought to be governed neither by the arithmetical ratio nor by the geometrical ratio if one wishes that the probability of error should never be above nor below a given fraction.

But what fraction ought to be determined upon? It is here that the arbitrariness begins and the tribunals offer in this regard the greatest variety. In the special tribunals where five of the eight votes suffice for the condemnation of the accused, the probability of the error to be feared in regard to justice of the judgment is $\frac{65}{256}$, or more than $\frac{1}{4}$. The magnitude of this fraction is dreadful; but that which ought to reassure us a little is the consideration that most frequently the judge who acquits an accused does not regard him as innocent; he pronounces solely that it is not attained by proofs sufficient for condemnation. One is especially reassured by the pity which nature has placed in the heart of man and which disposes the mind to see only with reluctance a culprit in the accused submitted to his judgment. This sentiment, more active in those who have not the habitude of criminal judgments, compensates for the inconveniences attached to the inexperience of the jurors. In a jury of twelve members, if the plurality demanded for the condemnation is eight of twelve votes, the probability of the error to be feared $\frac{1003}{8192}$, or a little more than one eighth, it is almost $\frac{1}{22}$ if this plurality consists of nine votes. In the case of unanimity the probability of the error to be feared is $\frac{1}{8192}$, that is to say, more than a thousand times less than in our juries. This supposes that the unanimity results only from proofs favorable or contrary to the accused; but motives that are entirely strange, ought oftentimes to concur in producing it, when it is imposed upon the jury as a necessary condition of its judgment. Then its decisions depending upon the temperament, the character, the habits of the jurors, and the circumstances in which they are placed, they are sometimes contrary to the decisions which the majority of the jury would have made if they had listened only to the proofs; this seems to me to be a great fault of this manner of judging.

The probability of the decision is too feeble in our juries, and I think that in order to give a sufficient guarantee to innocence, one ought to demand at least a plurality of nine votes in twelve.

CHAPTER XIV.

CONCERNING TABLES OF MORTALITY, AND OF MEAN DURATIONS OF LIFE, OF MARRIAGES, AND OF ASSOCIATIONS.

THE manner of preparing tables of mortality is very simple. One takes in the civil registers a great number of individuals whose birth and death are indicated. One determines how many of these individuals have died in the first year of their age, how many in the second year, and so on. It is concluded from these the number of individuals living at the commencement of each year, and this number is written in the table at the side of that which indicates the year. Thus one writes at the side of zero the number of births; at the side of the year 1 the number of infants who have attained one year; at the side of the year 2 the number of infants who have attained two years, and so on for the rest. But since in the first two years of life the mortality is very great, it is necessary for the sake of greater exactitude to indicate in this first age the number of survivors at the end of each half year.

If we divide the sum of the years of the life of all the individuals inscribed in a table of mortality by the number of these individuals we shall have the mean duration of life which corresponds to this table. For this, we will multiply by a half year the number of deaths in the first year, a number equal to the difference of the numbers of individuals inscribed at the side of the years 0 and 1. Their mortality being distributed over the entire year the mean duration of their life is only a half year. We will multiply by a year and a half the number of deaths in the second year; by two years and a half the number of deaths in the third year; and so on. The sum of these products divided by the number of births will be the mean duration of life. It is easy to conclude from this that we will obtain this duration, by making the sum of the numbers inscribed in the table at the side of each year, dividing it by the number of births and subtracting one half from the quotient, the year being taken as unity. The mean duration of life that remains, starting from any age, is determined in the same manner, working upon the number of individuals who have arrived at this age, as has just been done with the number of births. But it is not at the moment of birth that the mean duration of life is the greatest; it is when one has escaped the dangers of infancy and it is then about forty-three years. The probability of arriving at a certain age, starting from a given age is equal to the ratio of the two numbers of individuals indicated in the table at these two ages.

The precision of these results demands that for the formation of tables we should employ a very great number of births. Analysis gives then very simple formulæ for appreciating the probability that the numbers indicated in these tables will vary from the truth only within narrow limits. We see by these formulæ that the interval of the limits diminishes and that the probability increases in proportion as we take into consideration more births; so that the tables would represent exactly the true law of mortality if the number of births employed were infinite.

A table of mortality is then a table of the probability of human life. The ratio of the individuals inscribed at the side of each year to the number of births is the probability that a new birth will attain this year. As we estimate the value of hope by making a sum of the products of each benefit hoped for, by the probability of obtaining it, so we can equally evaluate the mean duration of life by adding the products of each year by half the sum of the probabilities of attaining the commencement and the end of it, which leads to the result found above. But this manner of viewing the mean duration of life has the advantage of showing that in a stationary population, that is to say, such that the number of births equals that of deaths, the mean duration of life is the ratio itself of the population to the annual births; for the population being supposed stationary, the number of individuals of an age comprised between two consecutive years of the table is equal to the number of annual births, multiplied by half the sum of the probabilities of attaining these years; the sum of all these products will be then the entire population. Now it is easy to see that this sum, divided by the number of annual births, coincides with the mean duration of life as we have just defined it.

It is easy by means of a table of mortality to form

the corresponding table of the population supposed to be stationary. For this we take the arithmetical means of the numbers of the table of mortality corresponding to the ages zero and one year, one and two years, two and three years, etc. The sum of all these means is the entire population; it is written at the side of the age zero. There is subtracted from this sum the first mean and the remainder is the number of individuals of one year and upwards; it is written at the side of the year 1. There is subtracted from this first remainder the second mean; this second remainder is the number of individuals of two years and upwards; it is written at the side of the year 2, and so on.

So many variable causes influence mortality that the tables which represent it ought to be changed according to place and time. The divers states of life offer in this regard appreciable differences relative to the fatigues and the dangers inseparable from each state and of which it is indispensable to keep account in the calculations founded upon the duration of life. But these differences have not been sufficiently observed. Some day they will be and then will be known what sacrifice of life each profession demands and one will profit by this knowledge to diminish the dangers.

The greater or less salubrity of the soil, its elevation, its temperature, the customs of the inhabitants, and the operations of governments have a considerable influence upon mortality. But it is always necessary to precede the investigation of the cause of the differences observed by that of the probability with which this cause is indicated. Thus the ratio of the population to annual births, which one has seen raised in France to twenty-eight and one third, is not equal to twenty-five in the ancient duchy of Milan. These ratios, both established upon a great number of births, do not permit of calling into question the existence among the Milanese of a special cause of mortality, which it is of moment for the government of our country to investigate and remove.

The ratio of the population to the births would increase again if we could diminish and remove certain dangerous and widely spread maladies. This has happily been done for the smallpox, at first by the inoculation of this disease, then in a manner much more advantageous, by the inoculation of vaccine, the inestimable discovery of Jenner, who has thereby become one of the greatest benefactors of humanity.

The smallpox has this in particular, namely, that the same individual is not twice affected by it, or at least such cases are so rare that they may be abstracted from the calculation. This malady, from which few escaped before the discovery of vaccine, is often fatal and causes the death of one seventh of those whom it attacks. Sometimes it is mild, and experience has taught that it can be given this latter character by inoculating it upon healthy persons, prepared for it by a proper diet and in a favorable season. Then the ratio of the individuals who die to the inoculated ones is not one three hundredth. This great advantage of inoculation, joined to those of not altering the appearance and of preserving from the grievous consequences which the natural smallpox often brings, caused it to be adopted by a great number of persons. The practice was strongly recommended, but it was strongly combated, as is nearly always the case in things subject to inconvenience. In the midst of this dispute Daniel Bernoulli proposed to submit to the calculus of probabilities the influence of inoculation upon the mean duration of life. Since precise data of the mortality produced by the smallpox at the various ages of life were lacking, he supposed that the danger of having this malady and that of dying of it are the same at every age. By means of these suppositions he succeeded by a delicate analysis in converting an ordinary table of mortality into that which would be used if smallpox did not exist, or if it caused the death of only a very small number of those affected, and he concludes from it that inoculation would augment by three years at least the mean duration of life, which appeared to him beyond doubt the advantage of this operation. D'Alembert attacked the analysis of Bernoulli: at first in regard to the uncertainty of his two hypotheses, then in regard to its insufficiency in this, that no comparison was made of the immediate danger, although very small, of dying of inoculation, to the very great but very remote danger of succumbing to natural smallpox. This consideration, which disappears when one considers a great number of individuals, is for this reason immaterial for governments and the advantages of inoculation for them still remain; but it is of great weight for the father of a family who must fear, in having his children inoculated, to see that one perish whom he holds most dear and to be the cause of it. Many parents were restrained by this fear, which the discovery of vaccine has happily dissipated. By one of those mysteries which nature offers to us so frequently, vaccine is a preventive of smallpox just as certain as variolar virus, and there is no danger at all; it does not expose to any malady and demands only very little care. Therefore the practice of it has spread quickly; and to render it universal it remains only to overcome the natural inertia of the people, against which it is necessary to strive continually, even when it is a question of their dearest interests.

The simplest means of calculating the advantage which the extinction of a malady would produce consists in determining by observation the number of individuals of a given age who die of it each year and subtracting this number from the number of deaths at the same age. The ratio of the difference to the total number of individuals of the given age would be the probability of dying in the year at this age if the malady did not exist. Making, then, a sum of these probabilities from birth up to any given age, and subtracting this sum from unity, the remainder will be the probability of living to that age corresponding to the extinction of the malady. The series of these probabilities will be the table of mortality relative to this hypothesis, and we may conclude from it, by what precedes, the mean duration of life. It is thus that Duvilard has found that the increase of the mean duration of life, due to inoculation with vaccine, is three years at the least. An increase so considerable would produce a very great increase in the population if the latter, for other reasons, were not restrained by the relative diminution of subsistences.

It is principally by the lack of subsistences that the progressive march of the population is arrested. In all kinds of animals and vegetables, nature tends without ceasing to augment the number of individuals until they are on a level of the means of subsistence. In the human race moral causes have a great influence upon the population. If easy clearings of the forest can furnish an abundant nourishment for new generations, the certainty of being able to support a numerous family encourages marriages and renders them more productive. Upon the same soil the population and the births ought to increase at the same time simultaneously in geometric progression. But when clearings become more difficult and more rare then the increase of population diminishes; it approaches continually the variable state of subsistences, making oscillations about it just as a pendulum whose periodicity is retarded by changing the point of suspension, oscillates about this point by virtue of its own weight. It is difficult to evaluate the *maximum* increase of the population; it appears after observations that in favorable circumstances the population of the human race would be doubled every fifteen years. We estimate that in North America the period of this doubling is twenty-two years. In this state of things, the population, births, marriages, mortality, all increase according to the same geometric progression of which we have the constant ratio of consecutive terms by the observation of annual births at two epochs.

By means of a table of mortality representing the probabilities of human life, we may determine the duration of marriages. Supposing in order to simplify the matter that the mortality is the same for the two sexes, we shall obtain the probability that the marriage will subsist one year, or two, or three, etc., by forming a series of fractions whose common denominator is the product of the two numbers of the table corresponding to the ages of the consorts, and whose numerators are the successive products of the numbers corresponding to these ages augmented by one, by two, by three, etc., years. The sum of these fractions augmented by one half will be the mean duration of marriage, the year being taken as unity. It is easy to extend the same rule to the mean duration of an association formed of three or of a greater number of individuals.

CHAPTER XV.

CONCERNING THE BENEFITS OF INSTITUTIONS WHICH DEPEND UPON THE PROBABILITY OF EVENTS.

LET us recall here what has been said in speaking of hope. It has been seen that in order to obtain the advantage which results from several simple events, of which the ones produce a benefit and the others a loss, it is necessary to add the products of the probability of each favorable event by the benefit which it procures, and subtract from their sum that of the products of the probability of each unfavorable event by the loss which is attached to it. But whatever may be the advantage expressed by the difference of these sums, a single event composed of these simple events does not guarantee against the fear of experiencing a loss. One imagines that this fear ought to decrease when one multiplies the compound event. The analysis of probabilities leads to this general theorem.

By the repetition of an advantageous event, simple or compound, the real benefit becomes more and more probable and increases without ceasing; it becomes certain in the hypothesis of an infinite number of repetitions; and dividing it by this number the quotient or the mean benefit of each event is the mathematical hope itself or the advantage relative to the event. It is the same with a loss which becomes certain in the long run, however small the disadvantage of the event may be.

This theorem upon benefits and losses is analogous to those which we have already given upon the ratios which are indicated by the indefinite repetition of events simple or compound; and, like them, it proves that regularity ends by establishing itself even in the things which are most subordinated to that which we name *hazard*.

When the events are in great number, analysis gives another very simple expression of the probability that the benefit will be comprised within determined limits. This is the expression which enters again into the general law of probability given above in speaking of the probabilities which result from the indefinite multiplication of events.

The stability of institutions which are based upon probabilities depends upon the truth of the preceding theorem. But in order that it may be applied to them it is necessary that those institutions should multiply these advantageous events for the sake of numerous things.

There have been based upon the probabilities of human life divers institutions, such as life annuities and tontines. The most general and the most simple method of calculating the benefits and the expenses of these institutions consists in reducing these to actual amounts. The annual interest of unity is that which is called *the rate of interest*. At the end of each year an amount acquires for a factor unity plus the rate of interest; it increases then according to a geometrical progression of which this factor is the ratio. Thus in the course of time it becomes immense. If, for example, the rate of interest is $\frac{1}{20}$ or five per cent, the capital doubles very nearly in fourteen years, quadruples in twenty-nine years, and in less than three centuries it becomes two million times larger.

An increase so prodigious has given birth to the idea of making use of it in order to pay off the public debt. One forms for this purpose a sinking fund to which is devoted an annual fund employed for the redemption of public bills and without ceasing increased by the interest of the bills redeemed. It is clear that in the long run this fund will absorb a great part of the national debt. If, when the needs of the State make a loan necessary, a part of this loan is devoted to the increasing of the annual sinking fund, the variation of public bills will be less; the confidence of the lenders and the probability of retiring without loss of capital loaned when one desires will be augmented and will render the conditions of the loan less onerous. Favorable experiences have fully confirmed these advantages. But the fidelity in engagements and the stability, so necessary to the success of such institutions, can be guaranteed only by a government in which the legislative power is divided among several independent powers. The confidence which the necessary coöperation of these powers inspires, doubles the strength of the State, and the sovereign himself gains then in legal power more than he loses in arbitrary power.

It results from that which precedes that the actual capital equivalent to a sum which is to be paid only after a certain number of years is equal to this sum multiplied by the probability that it will be paid at that time and divided by unity augmented by the rate of interest and raised to a power expressed by the number of these years.

It is easy to apply this principle to life annuities upon one or several persons, and to savings banks, and to assurance societies of any nature. Suppose that one proposes to form a table of life annuities according to a given table of mortality. A life annuity payable at the end of five years, for example, and reduced to an actual amount is, by this principle, equal to the product of the two following quantities, namely, the annuity divided by the fifth power of unity augmented by the rate of interest and the probability of paying it. This probability is the inverse ratio of the number of individuals inscribed in the table opposite to the age of that one who settles the annuity to the number inscribed opposite to this age augmented by five years. Forming, then, a series of fractions whose denominators are the products of the number of persons indicated in the table of mortality as living at the age of that one who settles the annuity, by the successive powers of unity augmented by the rate of interest, and whose numerators are the products of the annuity by the number of persons living at the same age augmented successively by one year, by two years, etc., the sum of these fractions will be the amount required for the life annuity at that age.

Let us suppose that a person wishes by means of a

life annuity to assure to his heirs an amount payable at the end of the year of his death. In order to determine the value of this annuity, one may imagine that the person borrows in life at a bank this capital and that he places it at perpetual interest in the same bank. It is clear that this same capital will be due by the bank to his heirs at the end of the year of his death; but he will have paid each year only the excess of the life interest over the perpetual interest. The table of life annuities will then show that which the person ought to pay annually to the bank in order to assure this capital after his death.

Maritime assurance, that against fire and storms, and generally all the institutions of this kind, are computed on the same principles. A merchant having vessels at sea wishes to assure their value and that of their cargoes against the dangers that they may run; in order to do this, he gives a sum to a company which becomes responsible to him for the estimated value of his cargoes and his vessels. The ratio of this value to the sum which ought to be given for the price of the assurance depends upon the dangers to which the vessels are exposed and can be appreciated only by numerous observations upon the fate of vessels which have sailed from port for the same destination.

If the persons assured should give to the assurance company only the sum indicated by the calculus of probabilities, this company would not be able to provide for the expenses of its institution; it is necessary then that they should pay a sum much greater than the cost of such insurance. What then is their advantage? It is here that the consideration of the moral disadvantage attached to an uncertainty becomes necessary. One conceives that the fairest game becomes, as has already been seen, disadvantageous, because the player exchanges a certain stake for an uncertain benefit; assurance by which one exchanges the uncertain for the certain ought to be advantageous. It is indeed this which results from the rule which we have given above for determining moral hope and by which one sees moreover how far the sacrifice may extend which ought to be made to the assurance company by reserving always a moral advantage. This company can then in procuring this advantage itself make a great benefit, if the number of the assured persons is very large, a condition necessary to its continued existence. Then its benefits become certain and the mathematical and moral hopes coincide; for analysis leads to this general theorem, namely, that if the expectations are very numerous the two hopes approach each other without ceasing and end by coinciding in the case of an infinite number.

We have said in speaking of mathematical and moral hopes that there is a moral advantage in distributing the risks of a benefit which one expects over several of its parts. Thus in order to send a sum of money to a distant part it is much better to send it on several vessels than to expose it on one. This one does by means of mutual assurances. If two persons, each having the same sum upon two different vessels which have sailed from the same port to the same destination, agree to divide equally all the money which may arrive, it is clear that by this agreement each of them divides equally between the two vessels the sum which he expects. Indeed this kind of assurance always leaves uncertainty as to the loss which one may fear. But this uncertainty diminishes in proportion as the number of policy-holders increases; the moral advantage increases more and more and ends by coinciding with the mathematical advantage, its natural limit. This renders the association of mutual assurances when it is very numerous more advantageous to the assured ones than the companies of assurance which, in proportion to the benefit that they give, give a moral advantage always inferior to the mathematical advantage. But the surveillance of their administration can balance the advantage of the mutual assurances. All these results are, as has already been seen, independent of the law which expresses the moral advantage.

One may look upon a free people as a great association whose members secure mutually their properties by supporting proportionally the charges of this guaranty. The confederation of several peoples would give to them advantages analogous to those which each individual enjoys in the society. A congress of their representatives would discuss objects of a utility common to all and without doubt the system of weights, measures, and moneys proposed by the French scientists would be adopted in this congress as one of the things most useful to commerical relations.

Among the institutions founded upon the probabilities of human life the better ones are those in which, by means of a light sacrifice of his revenue, one assures his existence and that of his family for a time when one ought to fear to be unable to satisfy their needs. As far as games are immoral, so far these institutions are advantageous to customs by favoring the strongest bents of our nature. The government ought then to encourage them and respect them in the vicissitudes of public fortune; since the hopes which they present look toward a distant future, they are able to prosper only when sheltered from all inquietude during their existence. It is an advantage that the institution of a representative government assures them.

Let us say a word about loans. It is clear that in order to borrow perpetually it is necessary to pay each year the product of the capital by the rate of interest. But one may wish to discharge this principal in equal payments made during a definite number of years, payments which are called *annuities* and whose value is obtained in this manner. Each annuity in order to be reduced at the actual moment ought to be divided by a power of unity augmented by the rate of interest equal to the number of years after which this annuity ought to be paid. Forming then a geometric progression whose first term is the annuity divided by unity augmented by the rate of interest, and whose last term is this annuity divided by the same quantity raised to a power equal to the number of years during which the payment should have been made, the sum of this progression will be equivalent to the capital borrowed, which will determine the value of the annuity. A sinking fund is at bottom only a means of converting into annuities a perpetual rent with the sole difference that in the case of a loan by annuities the interest is supposed constant, while the interest of funds acquired by the sinking fund is variable. If it were the same in both cases, the annuity corresponding to the funds acquired would be formed by these funds and from this annuity the State contributes annually to the sinking fund.

If one wishes to make a life loan it will be observed that the tables of life annuities give the capital required to constitute a life annuity at any age, a simple proportion will give the rent which one ought to pay to the individual from whom the capital is borrowed. From these principles all the possible kinds of loans may be calculated.

The principles which we have just expounded concerning the benefits and the losses of institutions may serve to determine the mean result of any number of observations already made, when one wishes to regard the deviations of the results corresponding to divers observations. Let us designate by x the correction of the least result and by x augmented successively by q, q', q'', etc., the corrections of the following results. Let us name e, e', e'', etc., the errors of the observations whose law of probability we will suppose known. Each observation being a function of the result, it is easy to see that by supposing the correction x of this result to be very small, the error e of the first observation will be equal to the product of x by a determined coefficient. Likewise the error e' of the second observation will be the product of the sum q plus x, by a determined coefficient, and so on. The probability of the error e being given by a known function, it will be expressed by the same function of the first of the preceding products. The probability of e' will be expressed by the same function of the second of these products, and so on of the others. The probability of the simultaneous existence of the errors e, e', e'', etc., will be then proportional to the product of these divers functions, a product which will be a function of x. This being granted, if one conceives a curve whose abscissa is x, and whose corresponding ordinate is this product, this curve will represent the probability of the divers values of x, whose limits will be determined by the limits of the errors e, e', e'', etc. Now let us designate by X the abscissa which it is necessary to choose; X diminished by x will be the error which would be committed if the abscissa x were the true correction. This error, multiplied by the probability of x or by the corresponding ordinate of the curve, will be the product of the loss by its probability, regarding, as one should, this error as a loss attached to the choice X. Multiplying this product by the differential of x the integral taken from the first extremity of the curve to X will be the disadvantage of X resulting from the values of x inferior to X. For the values of x superior to X, x less X would be the error of X if x were the true correction; the integral of the product of x by the corresponding ordinate of the curve and by the differential of x will be then the disadvantage of X resulting from the values x superior to x, this integral being taken from x equal to X up to the last extremity of the curve. Adding this disadvantage to the preceding one, the sum will be the disadvantage attached to the choice of X. This choice ought to be determined by the condition that this disadvantage be a *minimum;* and a very simple calculation shows that for this, X ought to be the abscissa whose ordinate divides the curve into two equal parts, so that it is thus probable that the true value of x falls on neither the one side nor the other of X.

Celebrated geometricians have chosen for X the most probable value of x and consequently that which corresponds to the largest ordinate of the curve; but the preceding value appears to me evidently that which the theory of probability indicates.

CHAPTER XVI.

CONCERNING ILLUSIONS IN THE ESTIMATION OF PROBABILITIES.

THE mind has its illusions as the sense of sight; and in the same manner that the sense of feeling corrects the latter, reflection and calculation correct the former. Probability based upon a daily experience, or exaggerated by fear and by hope, strikes us more than a superior probability but it is only a simple result of calculus. Thus we do not fear in return for small advantages to expose our life to dangers much less improbable than the drawing of a quint in the lottery of France; and yet no one would wish to procure for himself the same advantages with the certainty of losing his life if this quint should be drawn.

Our passions, our prejudices, and dominating opinions, by exaggerating the probabilities which are favorable to them and by attenuating the contrary probabilities, are the abundant sources of dangerous illusions.

Present evils and the cause which produced them effect us much more than the remembrance of evils produced by the contrary cause; they prevent us from appreciating with justice the inconveniences of the ones and the others, and the probability of the proper means to guard ourselves against them. It is this which leads alternately to despotism and to anarchy the people who are driven from the state of repose to which they never return except after long and cruel agitations.

This vivid impression which we receive from the presence of events, and which allows us scarcely to remark the contrary events observed by others, is a principal cause of error against which one cannot suffi-ciently guard himself.

It is principally at games of chance that a multitude of illusions support hope and sustain it against unfavor-able chances. The majority of those who play at lotteries do not know how many chances are to their advantage, how many are contrary to them. They see only the possibility by a small stake of gaining a considerable sum, and the projects which their imagi-nation brings forth, exaggerate to their eyes the probability of obtaining it; the poor man especially, excited by the desire of a better fate, risks at play his necessities by clinging to the most unfavorable com-binations which promise him a great benefit. All would be without doubt surprised by the immense number of stakes lost if they could know of them; but one takes care on the contrary to give to the winnings a great publicity, which becomes a new cause of excite-ment for this funereal play.

When a number in the lottery of France has not been drawn for a long time the crowd is eager to cover it with stakes. They judge since the number has not been drawn for a long time that it ought at the next drawing to be drawn in preference to others. So common an error appears to me to rest upon an illusion by which one is carried back involuntarily to the origin of events. It is, for example, very improbable that at the play of heads and tails one will throw heads ten times in succession. This improbability which strikes us indeed when it has happened nine times, leads us to believe that at the tenth throw tails will be thrown. But the past indicating in the coin a greater propensity for heads than for tails renders the first of the events more probable than the second; it increases as one has seen the probability of throwing heads at the following throw. A similar illusion persuades many people that one can certainly win in a lottery by placing each time upon the same number, until it is drawn, a stake whose product surpasses the sum of all the stakes. But even when similar speculations would not often be stopped by the impossibility of sustaining them they would not diminish the mathematical disadvantage of speculators and they would increase their moral disadvantage, since at each drawing they would risk a very large part of their fortune.

I have seen men, ardently desirous of having a son, who could learn only with anxiety of the births of boys in the month when they expected to become fathers. Imagining that the ratio of these births to those of girls ought to be the same at the end of each month, they judged that the boys already born would render more probable the births next of girls. Thus the extraction of a white ball from an urn which contains a limited number of white balls and of black balls increases the probability of extracting a black ball at the following drawing. But this ceases to take place when the number of balls in the urn is unlimited, as one must suppose in order to compare this case with that of births. If, in the course of a month, there were born many more boys than girls, one might suspect that toward the time of their conception a general cause had favored masculine conception, which would render more probable the birth next of a boy. The irregular events of nature are not exactly comparable to the drawing of the numbers of a lottery in which all the numbers are mixed at each drawing in such a manner as to render the chances of their drawing perfectly equal. The frequency of one of these events seems to indicate a cause slightly favoring it, which increases the probability of its next return, and its repetition prolonged for a long time, such as a long series of rainy days, may develop unknown causes for its change; so that at each expected event we are not, as at each drawing of a lottery, led back to the same state of indecision in regard to what ought to happen. But in proportion as the observation of these events is multiplied, the comparison of their results with those of lotteries becomes more exact.

By an illusion contrary to the preceding ones one seeks in the past drawings of the lottery of France the numbers most often drawn, in order to form combinations upon which one thinks to place the stake to advantage. But when the manner in which the mixing of the numbers in this lottery is considered, the past ought to have no influence upon the future. The very frequent drawings of a number are only the anomalies of chance; I have submitted several of them to calculation and have constantly found that they are included within the limits which the supposition of an equal possibility of the drawing of all the numbers allows us to admit without improbability.

In a long series of events of the same kind the single chances of hazard ought sometimes to offer the singular veins of good luck or bad luck which the majority of players do not fail to attribute to a kind of fatality. It happens often in games which depend at the same time upon hazard and upon the competency of the players, that that one who loses, troubled by his loss, seeks to repair it by hazardous throws which he would shun in another situation; thus he aggravates his own ill luck and prolongs its duration. It is then that prudence becomes necessary and that it is of importance to convince oneself that the moral disadvantage attached to unfavorable chances is increased by the ill luck itself.

The opinion that man has long been placed in the centre of the universe, considering himself the special object of the cares of nature, leads each individual to make himself the centre of a more or less extended sphere and to believe that hazard has preference for him. Sustained by this belief, players often risk considerable sums at games when they know that the chances are unfavorable. In the conduct of life a similar opinion may sometimes have advantages; but most often it leads to disastrous enterprises. Here as everywhere illusions are dangerous and truth alone is generally useful.

One of the great advantages of the calculus of probabilities is to teach us to distrust first opinions. As we recognize that they often deceive when they may be

submitted to calculus, we ought to conclude that in
other matters confidence should be given only after
extreme circumspection. Let us prove this by example.

An urn contains four balls, black and white, but which
are not all of the same color. One of these balls has
been drawn whose color is white and which has been
put back in the urn in order to proceed again to similar
drawings. One demands the probability of extracting
only black balls in the four following drawings.

If the white and black were in equal number this
probability would be the fourth power of the probability
$\frac{1}{2}$ of extracting a black ball at each drawing; it would
be then $\frac{1}{16}$. But the extraction of a white ball at the
first drawing indicates a superiority in the number of
white balls in the urn; for if one supposes in the urn
three white balls and one black the probability of
extracting a white ball is $\frac{3}{4}$; it is $\frac{2}{4}$ if one supposes two
white balls and two black; finally it is reduced to $\frac{1}{4}$ if
one supposes three black balls and one white. Following the principle of the probability of causes drawn
from events the probabilities of these three suppositions
are among themselves as the quantities $\frac{3}{4}, \frac{2}{4}, \frac{1}{4}$; they
are consequently equal to $\frac{3}{6}, \frac{2}{6}, \frac{1}{6}$. It is thus a bet of
5 against 1 that the number of black balls is inferior,
or at the most equal, to that of the white. It seems
then that after the extraction of a white ball at the first
drawing, the probability of extracting successively four
black balls ought to be less than in the case of the
equality of the colors or smaller than one sixteenth.
However, it is not, and it is found by a very simple
calculation that this probability is greater than one
fourteenth. Indeed it would be the fourth power

of $\frac{1}{4}$, of $\frac{2}{4}$, and of $\frac{3}{4}$ in the first, the second, and the third of the preceding suppositions concerning the colors of the balls in the urn. Multiplying respectively each power by the probability of the corresponding supposition, or by $\frac{3}{8}$, $\frac{2}{8}$, and $\frac{1}{8}$, the sum of the products will be the probability of extracting successively four black balls. One has thus for this probability $\frac{29}{384}$, a fraction greater than $\frac{1}{14}$. This paradox is explained by considering that the indication of the superiority of white balls over the black ones at the first drawing does not exclude at all the superiority of the black balls over the white ones, a superiority which excludes the supposition of the equality of the colors. But this superiority, though but slightly probable, ought to render the probability of drawing successively a given number of black balls greater than in this supposition if the number is considerable; and one has just seen that this commences when the given number is equal to four. Let us consider again an urn which contains several white and black balls. Let us suppose at first that there is only one white ball and one black. It is then an even bet that a white ball will be extracted in one drawing. But it seems for the equality of the bet that one who bets on extracting the white ball ought to have two drawings if the urn contains two black and one white, three drawings if it contains three black and one white, and so on; it is supposed that after each drawing the extracted ball is placed again in the urn.

We are convinced easily that this first idea is erroneous. Indeed in the case of two black and one white ball, the probability of extracting two black in two drawings is the second power of $\frac{2}{3}$ or $\frac{4}{9}$; but this probability added to that of drawing a white ball in two drawings is certainty or unity, since it is certain that two black balls or at least one white ball ought to be drawn; the probability in this last case is then $\frac{4}{5}$, a fraction greater than $\frac{1}{2}$. There would still be a greater advantage in the bet of drawing one white ball in five draws when the urn contains five black and one white ball; this bet is even advantageous in four drawings; it returns then to that of throwing six in four throws with a single die.

The Chevalier de Meré, who caused the invention of the calculus of probabilities by encouraging his friend Pascal, the great geometrician, to occupy himself with it, said to him "that he had found error in the numbers by this ratio. If we undertake to make six with one die there is an advantage in undertaking it in four throws, as 671 to 625. If we undertake to make two sixes with two dice, there is a disadvantage in undertaking in 24 throws. At least 24 is to 36, the number of the faces of the two dice, as 4 is to 6, the number of faces of one die." "This was," wrote Pascal to Fermat, "his great scandal which caused him to say boldly that the propositions were not constant and that arithmetic was demented. . . . He has a very good mind, but he is not a geometrician, which is, as you know, a great fault." The Chevalier de Meré, deceived by a false analogy, thought that in the case of the equality of bets the number of throws ought to increase in proportion to the number of all the chances possible, which is not exact, but which approaches exactness as this number becomes larger.

One has endeavored to explain the superiority of the

births of boys over those of girls by the general desire of fathers to have a son who would perpetuate the name. Thus by imagining an urn filled with an infinity of white and black balls in equal number, and supposing a great number of persons each of whom draws a ball from this urn and continues with the intention of stopping when he shall have extracted a white ball, one has believed that this intention ought to render the number of white balls extracted superior to that of the black ones. Indeed this intention gives necessarily after all the drawings a number of white balls equal to that of persons, and it is possible that these drawings would never lead a black ball. But it is easy to see that this first notion is only an illusion; for if one conceives that in the first drawing all the persons draw at once a ball from the urn, it is evident that their intention can have no influence upon the color of the balls which ought to appear at this drawing. Its unique effect will be to exclude from the second drawing the persons who shall have drawn a white one at the first. It is likewise apparent that the intention of the persons who shall take part in the new drawing will have no influence upon the color of the balls which shall be drawn, and that it will be the same at the following drawings. This intention will have no influence then upon the color of the balls extracted in the totality of drawings; it will, however, cause more or fewer to participate at each drawing. The ratio of the white balls extracted to the black ones will differ thus very little from unity. It follows that the number of persons being supposed very large, if observation gives between the colors extracted a ratio which differs sensibly from unity, it is very probable that the same difference is found between unity and the ratio of the white balls to the black contained in the urn.

I count again among illusions the application which Liebnitz and Daniel Bernoulli have made of the calculus of probabilities to the summation of series. If one reduces the fraction whose numerator is unity and whose denominator is unity plus a variable, in a series prescribed by the ratio to the powers of this variable, it is easy to see that in supposing the variable equal to unity the fraction becomes $\frac{1}{2}$, and the series becomes plus one, minus one, plus one, minus one, etc. In adding the first two terms, the second two, and so on, the series is transformed into another of which each term is zero. Grandi, an Italian Jesuit, concluded from this the possibility of the creation; because the series being always $\frac{1}{2}$, he saw this fraction spring from an infinity of zeros or from nothing. It was thus that Liebnitz believed he saw the image of creation in his binary arithmetic where he employed only the two characters, unity and zero. He imagined, since God can be represented by unity and nothing by zero, that the Supreme Being had drawn from nothing all beings, as unity with zero expresses all the numbers in this system of arithmetic. This idea was so pleasing to Liebnitz that he communicated it to the Jesuit Grimaldi, president of the tribunal of methematics in China, in the hope that this emblem of creation would convert to Christianity the emperor there who particularly loved the sciences. I report this incident only to show to what extent the prejudices of infancy can mislead the greatest men.

Liebnitz, always led by a singular and very loose metaphysics, considered that the series plus one, minus one, plus one, etc., becomes unity or zero according as one stops at a number of terms odd or even; and as in infinity there is no reason to prefer the even number to the odd, one ought following the rules of probability, to take the half of the results relative to these two kinds of numbers, and which are zero and unity, which gives $\frac{1}{2}$ for the value of the series. Daniel Bernoulli has since extended this reasoning to the summation of series formed from periodic terms. But all these series have no values properly speaking; they get them only in the case where their terms are multiplied by the successive powers of a variable less than unity. Then these series are always convergent, however small one supposes the difference of the variable from unity; and it is easy to demonstrate that the values assigned by Bernoulli, by virtue of the rule of probabilities, are the same values of the generative fraction of the series, when one supposes in these fractions the variable equal to unity. These values are again the limits which the series approach more and more, in proportion as the variable approaches unity. But when the variable is exactly equal to unity the series cease to be convergent; they have values only as far as one arrests them. The remarkable ratio of this application of the calculus of probabilities with the limits of the values of periodic series supposes that the terms of these series are multiplied by all the consecutive powers of the variable. But this series may result from the development of an infinity of different fractions in which this did not occur. Thus the series plus one, minus one, plus one, etc., may spring from the development of a fraction whose numerator is unity plus the variable, and whose denominator is this numerator augmented by the square of the variable. Supposing the variable equal to unity, this development changes, in the series proposed, and the generative fraction becomes equal to $\frac{2}{8}$; the rules of probabilities would give then a false result, which proves how dangerous it would be to employ similar reasoning, especially in the mathematical sciences, which ought to be especially distinguished by the rigor of their operations.

We are led naturally to believe that the order according to which we see things renewed upon the earth has existed from all times and will continue always. Indeed if the present state of the universe were exactly similar to the anterior state which has produced it, it would give birth in its turn to a similar state; the succession of these states would then be eternal. I have found by the application of analysis to the law of universal gravity that the movement of rotation and of revolution of the planets and satellites, and the position of the orbits and of their equators are subjected only to periodic inequalities. In comparing with ancient eclipses the theory of the secular equation of the moon I have found that since Hipparchus the duration of the day has not varied by the hundredth of a second, and that the mean temperature of the earth has not diminished the one-hundredth of a degree. Thus the stability of actual order appears established at the same time by theory and by observations. But this order is effected by divers causes which an attentive examination reveals, and which it is impossible to submit to calculus.

The actions of the ocean, of the atmosphere, and of meteors, of earthquakes, and the eruptions of volcanoes, agitate continually the surface of the earth and ought to effect in the long run great changes. The temperature of climates, the volume of the atmosphere, and the proportion of the gases which constitute it, may vary in an inappreciable manner. The instruments and the means suitable to determine these variations being new, observation has been unable up to this time to teach us anything in this regard. But it is hardly probable that the causes which absorb and renew the gases constituting the air maintain exactly their respective proportions. A long series of centuries will show the alterations which are experienced by all these elements so essential to the conservation of organized beings. Although historical monuments do not go back to a very great antiquity they offer us nevertheless sufficiently great changes which have come about by the slow and continued action of natural agents. Searching in the bowels of the earth one discovers numerous débris of former nature, entirely different from the present. Moreover, if the entire earth was in the beginning fluid, as everything appears to indicate, one imagines that in passing from that state to the one which it has now, its surface ought to have experienced prodigious changes. The heavens itself in spite of the order of its movements, is not unchangeable. The resistance of light and of other ethereal fluids, and the attraction of the stars ought, after a great number of centuries, to alter considerably the planetary movements. The variations already observed in the stars and in the form of the nebulæ give us a presentiment of those which time will develop in the system of these great bodies. One may represent the successive states of the universe by a curve, of which time would be the abscissa and of which the ordinates are the divers states. Scarcely knowing an element of this curve we are far from being able to go back to its origin; and if in order to satisfy the imagination, always restless from our ignorance of the cause of the phenomena which interest it, one ventures some conjectures it is wise to present them only with extreme reserve.

There exists in the estimation of probabilities a kind of illusions, which depending especially upon the laws of the intellectual organization demands, in order to secure oneself against them, a profound examination of these laws. The desire to penetrate into the future and the ratios of some remarkable events, to the predictions of astrologers, of diviners and soothsayers, to presentiments and dreams, to the numbers and the days reputed lucky or unlucky, have given birth to a multitude of prejudices still very widespread. One does not reflect upon the great number of non-coincidences which have made no impression or which are unknown. However, it is necessary to be acquainted with them in order to appreciate the probability of the causes to which the coincidences are attributed. This knowledge would confirm without doubt that which reason tells us in regard to these prejudices. Thus the philosopher of antiquity to whom is shown in a temple, in order to exalt the power of the god who is adored there, the *ex veto* of all those who after having invoked it were saved from shipwreck, presents an incident consonant with the calculus of probabilities, observing that he does not see inscribed the names of those who, in spite of this invocation, have perished. Cicero has refuted all these prejudices with much reason and eloquence in his *Treatise on Divination*, which he ends by a passage which I shall cite; for one loves to find again among the ancients the thunderbolts of reason, which, after having dissipated all the prejudices by its light, shall become the sole foundation of human institutions.

"It is necessary," says the Roman orator, "to reject divination by dreams and all similar prejudices. Widespread superstition has subjugated the majority of minds and has taken possession of the feebleness of men. It is this we have expounded in our books upon the nature of the gods and especially in this work, persuaded that we shall render a service to others and to ourselves if we succeed in destroying superstition. However (and I desire especially in this regard my thought be well comprehended), in destroying superstition I am far from wishing to disturb religion. Wisdom enjoins us to maintain the institutions and the ceremonies of our ancestors, touching the cult of the gods. Moreover, the beauty of the universe and the order of celestial things force us to recognize some superior nature which ought to be remarked and admired by the human race. But as far as it is proper to propagate religion, which is joined to the knowledge of nature, so far it is necessary to work toward the extirpation of superstition, for it torments one, importunes one, and pursues one continually and in all places.

If one consult a diviner or a soothsayer, if one immolates a victim, if one regards the flight of a bird, if one encounters a Chaldean or an aruspex, if it lightens, if it thunders, if the thunderbolt strikes, finally, if there is born or is manifested a kind of prodigy, things one of which ought often to happen, then superstition dominates and leaves no repose. Sleep itself, this refuge of mortals in their troubles and their labors, becomes by it a new source of inquietude and fear.''

All these prejudices and the terrors which they inspire are connected with physiological causes which continue sometimes to operate strongly after reason has disabused us of them. But the repetition of acts contrary to these prejudices can always destroy them.

CHAPTER XVII.

CONCERNING THE VARIOUS MEANS OF APPROACHING CERTAINTY.

INDUCTION, analogy, hypotheses founded upon facts and rectified continually by new observations, a happy tact given by nature and strengthened by numerous comparisons of its indications with experience, such are the principal means for arriving at truth.

If one considers a series of objects of the same nature one perceives among them and in their changes ratios which manifest themselves more and more in proportion as the series is prolonged, and which, extending and generalizing continually, lead finally to the principle from which they were derived. But these ratios are enveloped by so many strange circumstances that it requires great sagacity to disentangle them and to recur to this principle: it is in this that the true genius of sciences consists. Analysis and natural philosophy owe their most important discoveries to this fruitful means, which is called *induction*. Newton was indebted to it for his theorem of the binomial and the principle of universal gravity. It is difficult to appreciate the probability of the results of induction, which is

176

based upon this that the simplest ratios are the most common; this is verified in the formulæ of analysis and is found again in natural phenomena, in crystallization, and in chemical combinations. This simplicity of ratios will not appear astonishing if we consider that all the effects of nature are only mathematical results of a small number of immutable laws.

Yet induction, in leading to the discovery of the general principles of the sciences, does not suffice to establish them absolutely. It is always necessary to confirm them by demonstrations or by decisive experiences; for the history of the sciences shows us that induction has sometimes led to inexact results. I shall cite, for example, a theorem of Fermat in regard to prime numbers. This great geometrician, who had meditated profoundly upon this theorem, sought a formula which, containing only prime numbers, gave directly a prime number greater than any other number assignable. Induction led him to think that two, raised to a power which was itself a power of two, formed with unity a prime number. Thus, two raised to the square plus one, forms the prime number five; two raised to the second power of two, or sixteen, forms with one the prime number seventeen. He found that this was still true for the eighth and the sixteenth power of two augmented by unity; and this induction, based upon several arithmetical considerations, caused him to regard this result as general. However, he avowed that he had not demonstrated it. Indeed, Euler recognized that this does not hold for the thirty-second power of two, which, augmented by unity, gives 4,294,967,297, a number divisible by 641.

We judge by induction that if various events, movements, for example, appear constantly and have been long connected by a simple ratio, they will continue to be subjected to it; and we conclude from this, by the theory of probabilities, that this ratio is due, not to hazard, but to a regular cause. Thus the equality of the movements of the rotation and the revolution of the moon; that of the movements of the nodes of the orbit and of the lunar equator, and the coincidence of these nodes; the singular ratio of the movements of the first three satellites of Jupiter, according to which the mean longitude of the first satellite, less three times that of the second, plus two times that of the third, is equal to two right angles; the equality of the interval of the tides to that of the passage of the moon to the meridian; the return of the greatest tides with the syzygies, and of the smallest with the quadratures; all these things, which have been maintained since they were first observed, indicate with an extreme probability, the existence of constant causes which geometricians have happily succeeded in attaching to the law of universal gravity, and the knowledge of which renders certain the perpetuity of these ratios.

The chancellor Bacon, the eloquent promoter of the true philosophical method, has made a very strange misuse of induction in order to prove the immobility of the earth. He reasons thus in the *Novum Organum*, his finest work: " The movement of the stars from the orient to the occident increases in swiftness, in proportion to their distance from the earth. This movement is swiftest with the stars; it slackens a little with Saturn, a little more with Jupiter, and so on to

the moon and the highest comets. It is still perceptible in the atmosphere, especially between the tropics,
on account of the great circles which the molecules of
the air describe there; finally, it is almost inappreciable
with the ocean; it is then nil for the earth." But this
induction proves only that Saturn, and the stars which
are inferior to it, have their own movements, contrary
to the real or apparent movement which sweeps the
whole celestial sphere from the orient to the occident,
and that these movements appear slower with the more
remote stars, which is conformable to the laws of
optics. Bacon ought to have been struck by the
inconceivable swiftness which the stars require in order
to accomplish their diurnal revolution, if the earth is
immovable, and by the extreme simplicity with which
its rotation explains how bodies so distant, the ones
from the others, as the stars, the sun, the planets, and
the moon, all seem subjected to this revolution. As
to the ocean and to the atmosphere, he ought not to
compare their movement with that of the stars which
are detached from the earth; but since the air and the
sea make part of the terrestrial globe, they ought to
participate in its movement or in its repose. It is
singular that Bacon, carried to great prospects by his
genius, was not won over by the majestic idea which
the Copernican system of the universe offers. He was
able, however, to find in favor of that system, strong
analogies in the discoveries of Galileo, which were
continued by him. He has given for the search after
truth the precept, but not the example. But by
insisting, with all the force of reason and of eloquence,
upon the necessity of abandoning the insignificant

subtilities of the school, in order to apply oneself to observations and to experiences, and by indicating the true method of ascending to the general causes of phenomena, this great philosopher contributed to the immense strides which the human mind made in the grand century in which he terminated his career.

Analogy is based upon the probability, that similar things have causes of the same kind and produce the same effects. This probability increase as the similitude becomes more perfect. Thus we judge without doubt that beings provided with the same organs, doing the same things, experience the same sensations, and are moved by the same desires. The probability that the animals which resemble us have sensations analogous to ours, although a little inferior to that which is relative to individuals of our species, is still exceedingly great; and it has required all the influence of religious prejudices to make us think with some philosophers that animals are mere automatons. The probability of the existence of feeling decreases in the same proportion as the similitude of the organs with ours diminishes, but it is always very great, even with insects. In seeing those of the same species execute very complicated things exactly in the same manner from generation to generation, and without having learned them, one is led to believe that they act by a kind of affinity analogous to that which brings together the molecules of crystals, but which, together with the sensation attached to all animal organization, produces, with the regularity of chemical combinations, combinations that are much more singular; one might, perhaps, name this mingling of elective affinities and sensations *animal affinity.* Although there exists a great analogy between the organization of plants and that of animals, it does not seem to me sufficient to extend to vegetables the sense of feeling; but nothing authorizes us in denying it to them.

Since the sun brings forth, bythe beneficent action of its light and of its heat, the animals and plants which cover the earth, we judge by analogy that it produces similar effects upon the other planets; for it is not natural to think that the cause whose activity we see developed in so many ways should be sterile upon so great a planet as Jupiter, which, like the terrestrial globe, has its days, its nights, and its years, and upon which observations indicate changes which suppose very active forces. Yet this would be giving too great an extension to analogy to conclude from it the similitude of the inhabitants of the planets and of the earth. Man, made for the temperature which he enjoys, and for the element which he breathes, would not be able, according to all appearance, to live upon the other planets. But ought there not to be an infinity of organization relative to the various constitutions of the globes of this universe ? If the single difference of the elements and of the climates make so much variety in terrestrial productions, how much greater the difference ought to be among those of the various planets and of their satellites! The most active imagination can form no idea of it; but their existence is very probable.

We are led by a strong analogy to regard the stars as so many suns endowed, like ours, with an attractive power proportional to the mass and reciprocal to the square of the distances; for this power being demonstrated for all the bodies of the solar system, and for their smallest molecules, it appears to appertain to all matter.   Already the movements of the small stars, which have been called *double*, on account of their being binary, appear to indicate it; a century at most of precise observations, by verifying their movements of revolution, the ones about the others, will place beyond doubt their reciprocal attractions.

The analogy which leads us to make each star the centre of a planetary system is far less strong than the preceding one; but it acquires probability by the hypothesis which has been proposed in regard to the formation of the stars and of the sun; for in this hypothesis each star, having been like the sun, primitively environed by a vast atmosphere, it is natural to attribute to this atmosphere the same effects as to the solar atmosphere, and to suppose that it has produced, in condensing, planets and satellites.

A great number of discoveries in the sciences is due to analogy.   I shall cite as one of the most remarkable, the discovery of atmospheric electricity, to which one has been led by the analogy of electric phenomena with the effects of thunder.

The surest method which can guide us in the search for truth, consists in rising by induction from phenomena to laws and from laws to forces.   Laws are the ratios which connect particular phenomena together: when they have shown the general principle of the forces from which they are derived, one verifies it either by direct experiences, when this is possible, or by examination if it agrees with known phenomena; and if by a rigorous analysis we see them proceed from this principle, even in their small details, and if, moreover, they are quite varied and very numerous, then science acquires the highest degree of certainty and of perfection that it is able to attain. Such, astronomy has become by the discovery of universal gravity. The history of the sciences shows that the slow and laborious path of induction has not always been that of inventors. The imagination, impatient to arrive at the causes, takes pleasure in creating hypotheses, and often it changes the facts in order to adapt them to its work; then the hypotheses are dangerous. But when one regards them only as the means of connecting the phenomena in order to discover the laws; when, by refusing to attribute them to a reality, one rectifies them continually by new observations, they are able to lead to the veritable causes, or at least put us in a position to conclude from the phenomena observed those which given circumstances ought to produce.

If we should try all the hypotheses which can be formed in regard to the cause of phenomena we should arrive, by a process of exclusion, at the true one. This means has been employed with success; sometimes we have arrived at several hypotheses which explain equally well all the facts known, and among which scholars are divided, until decisive observations have made known the true one. Then it is interesting, for the history of the human mind, to return to these hypotheses, to see how they succeed in explaining a great number of facts, and to investigate the changes which they ought to undergo in order to agree with the history of nature. It is thus that the system of Ptolemy, which is only the realization of celestial appearances, is transformed into the hypothesis of the movement of the planets about the sun, by rendering equal and parallel to the solar orbit the circles and the epicycles which he causes to be described annually, and the magnitude of which he leaves undetermined. It suffices, then, in order to change this hypothesis into the true system of the world, to transport the apparent movement of the sun in a sense contrary to the earth.

It is almost always impossible to submit to calculus the probability of the results obtained by these various means; this is true likewise for historical facts. But the totality of the phenomena explained, or of the testimonies, is sometimes such that without being able to appreciate the probability we cannot reasonably permit ourselves any doubt in regard to them. In the other cases it is prudent to admit them only with great reserve.

CHAPTER XVIII.

HISTORICAL NOTICE CONCERNING THE CALCULUS OF PROBABILITIES.

LONG ago were determined, in the simplest games, the ratios of the chances which are favorable or unfavorable to the players; the stakes and the bets were regulated according to these ratios. But no one before Pascal and Fermat had given the principles and the methods for submitting this subject to calculus, and no one had solved the rather complicated questions of this kind. It is, then, to these two great geometricians that we must refer the first elements of the science of probabilities, the discovery of which can be ranked among the remarkable things which have rendered illustrious the seventeenth century—the century which has done the greatest honor to the human mind. The principal problem which they solved by different methods, consists, as we have seen, in distributing equitably the stake among the players, who are supposed to be equally skilful and who agree to stop the game before it is finished, the condition of play being that, in order to win the game, one must gain a given number of points different for each of the players. It is clear that the distribution should be made proportionally to the respective probabilities of the players of winning this game, the probabilities depending upon the numbers of points which are still lacking. The method of Pascal is very ingenious, and is at bottom only the equation of partial differences of this problem applied in determining the successive probabilities of the players, by going from the smallest numbers to the following ones. This method is limited to the case of two players; that of Fermat, based upon combinations, applies to any number of players. Pascal believed at first that it was, like his own, restricted to two players; this brougnt about between them a discussion, at the conclusion of which Pascal recognized the generality of the method of Fermat.

Huygens united the divers problems which had already been solved and added new ones in a little treatise, the first that has appeared on this subject and which has the title *De Ratiociniis in ludo aleæ*. Several geometricians have occupied themselves with the subject since: Hudde, the great pensionary, Witt in Holland, and Halley in England, applied calculus to the probabilities of human life, and Halley published in this field the first table of mortality. About the same time Jacques Bernoulli proposed to geometricians various problems of probability, of which he afterwards gave solutions. Finally he composed his beautiful work entitled *Ars conjectandi*, which appeared seven years after his death, which occurred in 1706. The science of probabilities is more profoundly investigated in this work than in that of Huygens. The author gives a general theory of combinations and series, and applies it to several difficult questions concerning hazards. This work is still remarkable on account of the justice and the cleverness of view, the employment of the formula of the binomial in this kind of questions, and by the demonstration of this theorem, namely, that in multiplying indefinitely the observations and the experiences, the ratio of the events of different natures approaches that of their respective probabilities in the limits whose interval becomes more and more narrow in proportion as they are multiplied, and become less than any assignable quantity. This theorem is very useful for obtaining by observations the laws and the causes of phenomena. Bernoulli attaches, with reason, a great importance to his demonstration, upon which he has said to have meditated for twenty years.

In the interval, from the death of Jacques Bernoulli to the publication of his work, Montmort and Moivre produced two treatises upon the calculus of probabilities. That of Montmort has the title *Essai sur les Jeux de hasard;* it contains numerous applications of this calculus to various games. The author has added in the second edition some letters in which Nicolas Bernoulli gives the ingenious solutions of several difficult problems. The treatise of Moivre, later than that of Montmort, appeared at first in the *Transactions philosophiques* of the year 1711. Then the author published it separately, and he has improved it successively in three editions. This work is principally based upon the formula of the binomial and the problems which it contains have, like their solutions, a grand generality. But its distinguishing feature is the theory of recurrent series and their use in this subject.   This theory is the integration of linear equations of finite differences with constant coefficients, which Moivre made in a very happy manner.

In his work, Moivre has taken up again the theory of Jacques Bernoulli in regard to the probability of results determined by a great number of observations. He does not content himself with showing, as Bernoulli does, that the ratio of the events which ought to occur approaches without ceasing that of their respective probabilities; but he gives besides an elegant and simple expression of the probability that the difference of these two ratios is contained within the given limits. For this purpose he determines the ratio of the greatest term of the development of a very high power of the binomial to the sum of all its terms, and the hyperbolic logarithm of the excess of this term above the terms adjacent to it.

The greatest term being then the product of a considerable number of factors, his numerical calculus becomes impracticable.   In order to obtain it by a convergent approximation, Moivre makes use of a theorem of Stirling in regard to the mean term of the binomial raised to a high power, a remarkable theorem, especially in this, that it introduces the square root of the ratio of the circumference to the radius in an expression which seemingly ought to be irrelevant to this transcendent.   Moreover, Moivre was greatly struck by this result, which Stirling had deduced from the expression of the circumference in infinite products; Wallis had arrived at this expression by a singlar analysis which contains the germ of the very curious and useful theory of definite intergrals.

Many scholars, among whom one ought to name Deparcieux, Kersseboom, Wargentin, Dupré de Saint-Maure, Simpson, Sussmilch, Messène, Moheau, Price, Bailey, and Duvillard, have collected a great amount of precise data in regard to population, births, marriages, and mortality. They have given formulæ and tables relative to life annuities, tontines, assurances, etc. But in this short notice I can only indicate these useful works in order to adhere to original ideas. Of this number special mention is due to the mathematical and moral hopes and to the ingenious principle which Daniel Bernoulli has given for submitting the latter to analysis. Such is again the happy application which he has made of the calculus of probabilities to inoculation. One ought especially to include, in the number of these original ideas, direct consideration of the possibility of events drawn from events observed. Jacques Bernoulli and Moivre supposed these possibilities known, and they sought the probability that the result of future experiences will more and more nearly represent them. Bayes, in the *Transactions philosophiques* of the year 1763, sought directly the probability that the possibilities indicated by past experiences are comprised within given limits; and he has arrived at this in a refined and very ingenious manner, although a little perplexing. This subject is connected with the theory of the probability of causes and future events, concluded from events observed. Some years later I expounded the principles of this theory with a remark as to the influence of the inequalities which may exist among the chances which are supposed to be equal. Although it is not known which of the simple events these inequalities favor, nevertheless this ignorance itself often increases the probability of compound events.

In generalizing analysis and the problems concerning probabilities, I was led to the calculus of partial finite differences, which Lagrange has since treated by a very simple method, elegant applications of which he has used in this kind of problems. The theory of generative functions which I published about the same time includes these subjects among those it embraces, and is adapted of itself and with the greatest generality to the most difficult questions of probability. It determines again, by very convergent approximations, the values of the functions composed of a great number of terms and factors; and in showing that the square root of the ratio of the circumference to the radius enters most frequently into these values, it shows that an infinity of other transcendents may be introduced.

Testimonies, votes, and the decisions of electoral and deliberative assemblies, and the judgments of tribunals, have been submitted likewise to the calculus of probabilities. So many passions, divers interests, and circumstances complicate the questions relative to the subjects, that they are almost always insoluble. But the solution of very simple problems which have a great analogy with them, may often shed upon difficult and important questions great light, which the surety of calculus renders always preferable to the most specious reasonings.

One of the most interesting applications of the calculus of probabilities concerns the mean values which must be chosen among the results of observations. Many geometricians have studied the subject, and Lagrange has published in the *Mémoires de Turin* a beautiful method for determining these mean values when the law of the errors of the observations is known. I have given for the same purpose a method based upon a singular contrivance which may be employed with advantage in other questions of analysis; and this, by permitting indefinite extension in the whole course of a long calculation of the functions which ought to be limited by the nature of the problem, indicates the modifications which each term of the final result ought to receive by virtue of these limitations. It has already been seen that each observation furnishes an equation of condition of the first degree, which may always be disposed of in such a manner that all its terms be in the first member, the second being zero. The use of these equations is one of the principal causes of the great precision of our astronomical tables, because an immense number of excellent observations has thus been made to concur in determining their elements. When there is only one element to be determined Côtes prescribed that the equations of condition should be prepared in such a manner that the coefficient of the unknown element be positive in each of them; and that all these equations should be added in order to form a final equation, whence is derived the value of this element. The rule of Côtes was followed by all calculators, but since he failed to determine several elements, there was no fixed rule for combining the equations of condition in such a manner as to obtain the necessary final equations; but one chose for each element the observations most suitable to determine it.   It was in order to obviate these gropings that Legendre and Gauss concluded to add the squares of the first members of the equations of condition, and to render the sum a minimum, by varying each unknown element; by this means is obtained directly as many final equations as there are elements. But do the values determined by these equations merit the preference over all those which may be obtained by other means ?   This question, the calculus of probabilities alone was able to answer.   I applied it, then, to this subject, and obtained by a delicate analysis a rule which includes the preceding method, and which adds to the advantage of giving, by a regular process, the desired elements that of obtaining them with the greatest show of evidence from the totality of observations, and of determining the values which leave only the smallest possible errors to be feared.

However, we have only an imperfect knowledge of the results obtained, as long as the law of the errors of which they are susceptible is unknown; we must be able to assign the probability that these errors are contained within given limits, which amounts to determining that which I have called the *weight* of a result. Analysis leads to general and simple formulæ for this purpose.   I have applied this analysis to the results of geodetic observations.   The general problem consists in determining the probabilities that the values of one or of several linear functions of the errors of a very great number of observations are contained within any limits.

The law of the possibility of the errors of observations introduces into the expressions of these probabilities a constant, whose value seems to require the knowledge of this law, which is almost always unknown. Happily this constant can be determined from the observations.

In the investigation of astronomical elements it is given by the sum of the squares of the differences between each observation and the calculated one. The errors equally probable being proportional to the square root of this sum, one can, by the comparison of these squares, appreciate the relative exactitude of the different tables of the same star. In geodetic operations these squares are replaced by the squares of the errors of the sums observed of the three angles of each triangle. The comparison of the squares of these errors will enable us to judge of the relative precision of the instruments with which the angles have been measured. By this comparison is seen the advantage of the repeating circle over the instruments which it has replaced in geodesy.

There often exists in the observations many sources of errors: thus the positions of the stars being determined by means of the meridian telescope and of the circle, both susceptible of errors whose law of probability ought not to be supposed the same, the elements that are deduced from these positions are affected by these errors. The equations of condition, which are made to obtain these elements, contain the errors of each instrument and they have various coefficients. The most advantageous system of factors by which these equations ought to be multiplied respectively, in

order to obtain, by the union of the products, as many
final equations as there are elements to be determined,
is no longer that of the coefficients of the elements in
each equation of condition. The analysis which I have
used leads easily, whatever the number of the sources
of error may be, to the system of factors which gives
the most advantageous results, or those in which the
same error is less probable than in any other system.
The same analysis determines the laws of probability
of the errors of these results. These formulæ contain
as many unknown constants as there are sources of
error, and they depend upon the laws of probability of
these errors. It has been seen that, in the case of a
single source, this constant can be determined by
forming the sum of the squares of the residuals of each
equation of condition, when the values found for these
elements have been substituted. A similar process
generally gives values of these constants, whatever
their number may be, which completes the application
of the calculus of probabilities to the results of observations.

I ought to make here an important remark. The
small uncertainty that the observations, when they are
not numerous, leave in regard to the values of the
constants of which I have just spoken, renders a little
uncertain the probabilities determined by analysis.
But it almost always suffices to know if the probability,
that the errors of the results obtained are comprised
within narrow limits, approaches closely to unity; and
when it is not, it suffices to know up to what point the
observations should be multiplied, in order to obtain a
probability such that no reasonable doubt remains in

regard to the correctness of the results. The analytic formulæ of probabilities satisfy perfectly this requirement; and in this connection they may be viewed as the necessary complement of the sciences, based upon a totality of observations susceptible of error. They are likewise indispensable in solving a great number of problems in the natural and moral sciences. The regular causes of phenomena are most frequently either unknown, or too complicated to be submitted to calculus; again, their action is often disturbed by accidental and irregular causes; but its impression always remains in the events produced by all these causes, and it leads to modifications which only a long series of observations can determine. The analysis of probabilities develops these modifications; it assigns the probability of their causes and it indicates the means of continually increasing this probability. Thus in the midst of the irregular causes which disturb the atmosphere, the periodic changes of solar heat, from day to night, and from winter to summer, produce in the pressure of this great fluid mass and in the corresponding height of the barometer, the diurnal and annual oscillations; and numerous barometric observations have revealed the former with a probability at least equal to that of the facts which we regard as certain. Thus it is again that the series of historical events shows us the constant action of the great principles of ethics in the midst of the passions and the various interests which disturb societies in every way. It is remarkable that a science, which commenced with the consideration of games of chance, should be elevated to the rank of the most important subjects of human knowlegdge.

I have collected all these methods in my *Théorie analytique des Probabilités*, in which I have proposed to expound in the most general manner the principles and the analysis of the calculus of probabilities, likewise the solutions of the most interesting and most difficult problems which calculus presents.

It is seen in this essay that the theory of probabilities is at bottom only common sense reduced to calculus; it makes us appreciate with exactitude that which exact minds feel by a sort of instinct without being able ofttimes to give a reason for it.   It leaves no arbitrariness in the choice of opinions and sides to be taken; and by its use can always be determined the most advantageous choice.   Thereby it supplements most happily the ignorance and the weakness of the human mind.   If we consider the analytical methods to which this theory has given birth; the truth of the principles which serve as a basis; the fine and delicate logic which their employment in the solution of problems requires; the establishments of public utility which rest upon it; the extension which it has received and which it can still receive by its application to the most important questions of natural philosophy and the moral science; if we consider again that, even in the things which cannot be submitted to calculus, it gives the surest hints which can guide us in our judgments, and that it teaches us to avoid the illusions which ofttimes confuse us, then we shall see that there is no science more worthy of our meditations, and that no more useful one could be incorporated in the system of public instruction.

A CATALOG OF SELECTED
DOVER BOOKS
IN ALL FIELDS OF INTEREST

A CATALOG OF SELECTED DOVER
BOOKS IN ALL FIELDS OF INTEREST

CONCERNING THE SPIRITUAL IN ART, Wassily Kandinsky. Pioneering work by father of abstract art. Thoughts on color theory, nature of art. Analysis of earlier masters. 12 illustrations. 80pp. of text. 5⅜ x 8½. 23411-8 Pa. $4.95

ANIMALS: 1,419 Copyright-Free Illustrations of Mammals, Birds, Fish, Insects, etc., Jim Harter (ed.). Clear wood engravings present, in extremely lifelike poses, over 1,000 species of animals. One of the most extensive pictorial sourcebooks of its kind. Captions. Index. 284pp. 9 x 12. 23766-4 Pa. $14.95

CELTIC ART: The Methods of Construction, George Bain. Simple geometric techniques for making Celtic interlacements, spirals, Kells-type initials, animals, humans, etc. Over 500 illustrations. 160pp. 9 x 12. (USO) 22923-8 Pa. $9.95

AN ATLAS OF ANATOMY FOR ARTISTS, Fritz Schider. Most thorough reference work on art anatomy in the world. Hundreds of illustrations, including selections from works by Vesalius, Leonardo, Goya, Ingres, Michelangelo, others. 593 illustrations. 192pp. 7⅛ x 10¼. 20241-0 Pa. $9.95

CELTIC HAND STROKE-BY-STROKE (Irish Half-Uncial from "The Book of Kells"): An Arthur Baker Calligraphy Manual, Arthur Baker. Complete guide to creating each letter of the alphabet in distinctive Celtic manner. Covers hand position, strokes, pens, inks, paper, more. Illustrated. 48pp. 8¼ x 11. 24336-2 Pa. $3.95

EASY ORIGAMI, John Montroll. Charming collection of 32 projects (hat, cup, pelican, piano, swan, many more) specially designed for the novice origami hobbyist. Clearly illustrated easy-to-follow instructions insure that even beginning papercrafters will achieve successful results. 48pp. 8¼ x 11. 27298-2 Pa. $3.50

THE COMPLETE BOOK OF BIRDHOUSE CONSTRUCTION FOR WOODWORKERS, Scott D. Campbell. Detailed instructions, illustrations, tables. Also data on bird habitat and instinct patterns. Bibliography. 3 tables. 63 illustrations in 15 figures. 48pp. 5¼ x 8½. 24407-5 Pa. $2.50

BLOOMINGDALE'S ILLUSTRATED 1886 CATALOG: Fashions, Dry Goods and Housewares, Bloomingdale Brothers. Famed merchants' extremely rare catalog depicting about 1,700 products: clothing, housewares, firearms, dry goods, jewelry, more. Invaluable for dating, identifying vintage items. Also, copyright-free graphics for artists, designers. Co-published with Henry Ford Museum & Greenfield Village. 160pp. 8¼ x 11. 25780-0 Pa. $10.95

HISTORIC COSTUME IN PICTURES, Braun & Schneider. Over 1,450 costumed figures in clearly detailed engravings–from dawn of civilization to end of 19th century. Captions. Many folk costumes. 256pp. 8⅜ x 11¾. 23150-X Pa. $12.95

CATALOG OF DOVER BOOKS

STICKLEY CRAFTSMAN FURNITURE CATALOGS, Gustav Stickley and L. & J. G. Stickley. Beautiful, functional furniture in two authentic catalogs from 1910. 594 illustrations, including 277 photos, show settles, rockers, armchairs, reclining chairs, bookcases, desks, tables. 183pp. 6½ x 9¼. 23838-5 Pa. $11.95

AMERICAN LOCOMOTIVES IN HISTORIC PHOTOGRAPHS: 1858 to 1949, Ron Ziel (ed.). A rare collection of 126 meticulously detailed official photographs, called "builder portraits," of American locomotives that majestically chronicle the rise of steam locomotive power in America. Introduction. Detailed captions. xi + 129pp. 9 x 12. 27393-8 Pa. $13.95

AMERICA'S LIGHTHOUSES: An Illustrated History, Francis Ross Holland, Jr. Delightfully written, profusely illustrated fact-filled survey of over 200 American lighthouses since 1716. History, anecdotes, technological advances, more. 240pp. 8 x 10¾. 25576-X Pa. $12.95

TOWARDS A NEW ARCHITECTURE, Le Corbusier. Pioneering manifesto by founder of "International School." Technical and aesthetic theories, views of industry, economics, relation of form to function, "mass-production split" and much more. Profusely illustrated. 320pp. 6⅛ x 9¼. (USO) 25023-7 Pa. $9.95

HOW THE OTHER HALF LIVES, Jacob Riis. Famous journalistic record, exposing poverty and degradation of New York slums around 1900, by major social reformer. 100 striking and influential photographs. 233pp. 10 x 7⅞. 22012-5 Pa. $11.95

FRUIT KEY AND TWIG KEY TO TREES AND SHRUBS, William M. Harlow. One of the handiest and most widely used identification aids. Fruit key covers 120 deciduous and evergreen species; twig key 160 deciduous species. Easily used. Over 300 photographs. 126pp. 5⅜ x 8½. 20511-8 Pa. $3.95

COMMON BIRD SONGS, Dr. Donald J. Borror. Songs of 60 most common U.S. birds: robins, sparrows, cardinals, bluejays, finches, more–arranged in order of increasing complexity. Up to 9 variations of songs of each species. Cassette and manual 99911-4 $8.95

ORCHIDS AS HOUSE PLANTS, Rebecca Tyson Northen. Grow cattleyas and many other kinds of orchids–in a window, in a case, or under artificial light. 63 illustrations. 148pp. 5⅜ x 8½. 23261-1 Pa. $5.95

MONSTER MAZES, Dave Phillips. Masterful mazes at four levels of difficulty. Avoid deadly perils and evil creatures to find magical treasures. Solutions for all 32 exciting illustrated puzzles. 48pp. 8¼ x 11. 26005-4 Pa. $2.95

MOZART'S DON GIOVANNI (DOVER OPERA LIBRETTO SERIES), Wolfgang Amadeus Mozart. Introduced and translated by Ellen H. Bleiler. Standard Italian libretto, with complete English translation. Convenient and thoroughly portable–an ideal companion for reading along with a recording or the performance itself. Introduction. List of characters. Plot summary. 121pp. 5¼ x 8½. 24944-1 Pa. $3.95

TECHNICAL MANUAL AND DICTIONARY OF CLASSICAL BALLET, Gail Grant. Defines, explains, comments on steps, movements, poses and concepts. 15-page pictorial section. Basic book for student, viewer. 127pp. 5⅜ x 8½. 21843-0 Pa. $4.95

THE CLARINET AND CLARINET PLAYING, David Pino. Lively, comprehensive work features suggestions about technique, musicianship, and musical interpretation, as well as guidelines for teaching, making your own reeds, and preparing for public performance. Includes an intriguing look at clarinet history. "A godsend," The Clarinet, Journal of the International Clarinet Society. Appendixes. 7 illus. 320pp. 5⅜ x 8½. 40270-3 Pa. $9.95

HOLLYWOOD GLAMOR PORTRAITS, John Kobal (ed.). 145 photos from 1926-49. Harlow, Gable, Bogart, Bacall; 94 stars in all. Full background on photographers, technical aspects. 160pp. 8⅜ x 11¼. 23352-9 Pa. $12.95

THE ANNOTATED CASEY AT THE BAT: A Collection of Ballads about the Mighty Casey/Third, Revised Edition, Martin Gardner (ed.). Amusing sequels and parodies of one of America's best-loved poems: Casey's Revenge, Why Casey Whiffed, Casey's Sister at the Bat, others. 256pp. 5⅜ x 8½. 28598-7 Pa. $8.95

THE RAVEN AND OTHER FAVORITE POEMS, Edgar Allan Poe. Over 40 of the author's most memorable poems: "The Bells," "Ulalume," "Israfel," "To Helen," "The Conqueror Worm," "Eldorado," "Annabel Lee," many more. Alphabetic lists of titles and first lines. 64pp. 5³⁄₁₆ x 8¼. 26685-0 Pa. $1.00

PERSONAL MEMOIRS OF U. S. GRANT, Ulysses Simpson Grant. Intelligent, deeply moving firsthand account of Civil War campaigns, considered by many the finest military memoirs ever written. Includes letters, historic photographs, maps and more. 528pp. 6⅛ x 9¼. 28587-1 Pa. $12.95

ANCIENT EGYPTIAN MATERIALS AND INDUSTRIES, A. Lucas and J. Harris. Fascinating, comprehensive, thoroughly documented text describes this ancient civilization's vast resources and the processes that incorporated them in daily life, including the use of animal products, building materials, cosmetics, perfumes and incense, fibers, glazed ware, glass and its manufacture, materials used in the mummification process, and much more. 544pp. 6⅛ x 9¼. (USO) 40446-3 Pa. $16.95

RUSSIAN STORIES/PYCCKNE PACCKA3bl: A Dual-Language Book, edited by Gleb Struve. Twelve tales by such masters as Chekhov, Tolstoy, Dostoevsky, Pushkin, others. Excellent word-for-word English translations on facing pages, plus teaching and study aids, Russian/English vocabulary, biographical/critical introductions, more. 416pp. 5⅜ x 8½. 26244-8 Pa. $9.95

PHILADELPHIA THEN AND NOW: 60 Sites Photographed in the Past and Present, Kenneth Finkel and Susan Oyama. Rare photographs of City Hall, Logan Square, Independence Hall, Betsy Ross House, other landmarks juxtaposed with contemporary views. Captures changing face of historic city. Introduction. Captions. 128pp. 8¼ x 11. 25790-8 Pa. $9.95

AIA ARCHITECTURAL GUIDE TO NASSAU AND SUFFOLK COUNTIES, LONG ISLAND, The American Institute of Architects, Long Island Chapter, and the Society for the Preservation of Long Island Antiquities. Comprehensive, well-researched and generously illustrated volume brings to life over three centuries of Long Island's great architectural heritage. More than 240 photographs with authoritative, extensively detailed captions. 176pp. 8¼ x 11. 26946-9 Pa. $14.95

NORTH AMERICAN INDIAN LIFE: Customs and Traditions of 23 Tribes, Elsie Clews Parsons (ed.). 27 fictionalized essays by noted anthropologists examine religion, customs, government, additional facets of life among the Winnebago, Crow, Zuni, Eskimo, other tribes. 480pp. 6⅛ x 9¼. 27377-6 Pa. $10.95

FRANK LLOYD WRIGHT'S DANA HOUSE, Donald Hoffmann. Pictorial essay of residential masterpiece with over 160 interior and exterior photos, plans, elevations, sketches and studies. 128pp. 9¼ x 10¾. 29120-0 Pa. $12.95

THE MALE AND FEMALE FIGURE IN MOTION: 60 Classic Photographic Sequences, Eadweard Muybridge. 60 true-action photographs of men and women walking, running, climbing, bending, turning, etc., reproduced from rare 19th-century masterpiece. vi + 121pp. 9 x 12. 24745-7 Pa. $10.95

1001 QUESTIONS ANSWERED ABOUT THE SEASHORE, N. J. Berrill and Jacquelyn Berrill. Queries answered about dolphins, sea snails, sponges, starfish, fishes, shore birds, many others. Covers appearance, breeding, growth, feeding, much more. 305pp. 5¼ x 8¼. 23366-9 Pa. $9.95

ATTRACTING BIRDS TO YOUR YARD, William J. Weber. Easy-to-follow guide offers advice on how to attract the greatest diversity of birds: birdhouses, feeders, water and waterers, much more. 96pp. 5³⁄₁₆ x 8¼. 28927-3 Pa. $2.50

MEDICINAL AND OTHER USES OF NORTH AMERICAN PLANTS: A Historical Survey with Special Reference to the Eastern Indian Tribes, Charlotte Erichsen-Brown. Chronological historical citations document 500 years of usage of plants, trees, shrubs native to eastern Canada, northeastern U.S. Also complete identifying information. 343 illustrations. 544pp. 6½ x 9¼. 25951-X Pa. $12.95

STORYBOOK MAZES, Dave Phillips. 23 stories and mazes on two-page spreads: Wizard of Oz, Treasure Island, Robin Hood, etc. Solutions. 64pp. 8¼ x 11. 23628-5 Pa. $2.95

AMERICAN NEGRO SONGS: 230 Folk Songs and Spirituals, Religious and Secular, John W. Work. This authoritative study traces the African influences of songs sung and played by black Americans at work, in church, and as entertainment. The author discusses the lyric significance of such songs as "Swing Low, Sweet Chariot," "John Henry," and others and offers the words and music for 230 songs. Bibliography. Index of Song Titles. 272pp. 6½ x 9¼. 40271-1 Pa. $9.95

MOVIE-STAR PORTRAITS OF THE FORTIES, John Kobal (ed.). 163 glamor, studio photos of 106 stars of the 1940s: Rita Hayworth, Ava Gardner, Marlon Brando, Clark Gable, many more. 176pp. 8⅜ x 11¼. 23546-7 Pa. $14.95

BENCHLEY LOST AND FOUND, Robert Benchley. Finest humor from early 30s, about pet peeves, child psychologists, post office and others. Mostly unavailable elsewhere. 73 illustrations by Peter Arno and others. 183pp. 5⅜ x 8½. 22410-4 Pa. $6.95

YEKL and THE IMPORTED BRIDEGROOM AND OTHER STORIES OF YIDDISH NEW YORK, Abraham Cahan. Film Hester Street based on Yekl (1896). Novel, other stories among first about Jewish immigrants on N.Y.'s East Side. 240pp. 5⅜ x 8½. 22427-9 Pa. $6.95

SELECTED POEMS, Walt Whitman. Generous sampling from *Leaves of Grass*. Twenty-four poems include "I Hear America Singing," "Song of the Open Road," "I Sing the Body Electric," "When Lilacs Last in the Dooryard Bloom'd," "O Captain! My Captain!"—all reprinted from an authoritative edition. Lists of titles and first lines. 128pp. 5³⁄₁₆ x 8¼. 26878-0 Pa. $1.00

THE BEST TALES OF HOFFMANN, E. T. A. Hoffmann. 10 of Hoffmann's most important stories: "Nutcracker and the King of Mice," "The Golden Flowerpot," etc. 458pp. 5⅜ x 8½. 21793-0 Pa. $9.95

FROM FETISH TO GOD IN ANCIENT EGYPT, E. A. Wallis Budge. Rich detailed survey of Egyptian conception of "God" and gods, magic, cult of animals, Osiris, more. Also, superb English translations of hymns and legends. 240 illustrations. 545pp. 5⅜ x 8½. 25803-3 Pa. $13.95

FRENCH STORIES/CONTES FRANÇAIS: A Dual-Language Book, Wallace Fowlie. Ten stories by French masters, Voltaire to Camus: "Micromegas" by Voltaire; "The Atheist's Mass" by Balzac; "Minuet" by de Maupassant; "The Guest" by Camus, six more. Excellent English translations on facing pages. Also French-English vocabulary list, exercises, more. 352pp. 5⅜ x 8½. 26443-2 Pa. $9.95

CHICAGO AT THE TURN OF THE CENTURY IN PHOTOGRAPHS: 122 Historic Views from the Collections of the Chicago Historical Society, Larry A. Viskochil. Rare large-format prints offer detailed views of City Hall, State Street, the Loop, Hull House, Union Station, many other landmarks, circa 1904-1913. Introduction. Captions. Maps. 144pp. 9⅜ x 12¼. 24656-6 Pa. $12.95

OLD BROOKLYN IN EARLY PHOTOGRAPHS, 1865-1929, William Lee Younger. Luna Park, Gravesend race track, construction of Grand Army Plaza, moving of Hotel Brighton, etc. 157 previously unpublished photographs. 165pp. 8⅞ x 11¾. 23587-4 Pa. $13.95

THE MYTHS OF THE NORTH AMERICAN INDIANS, Lewis Spence. Rich anthology of the myths and legends of the Algonquins, Iroquois, Pawnees and Sioux, prefaced by an extensive historical and ethnological commentary. 36 illustrations. 480pp. 5⅜ x 8½. 25967-6 Pa. $10.95

AN ENCYCLOPEDIA OF BATTLES: Accounts of Over 1,560 Battles from 1479 B.C. to the Present, David Eggenberger. Essential details of every major battle in recorded history from the first battle of Megiddo in 1479 B.C. to Grenada in 1984. List of Battle Maps. New Appendix covering the years 1967-1984. Index. 99 illustrations. 544pp. 6½ x 9¼. 24913-1 Pa. $16.95

SAILING ALONE AROUND THE WORLD, Captain Joshua Slocum. First man to sail around the world, alone, in small boat. One of great feats of seamanship told in delightful manner. 67 illustrations. 294pp. 5⅜ x 8½. 20326-3 Pa. $6.95

ANARCHISM AND OTHER ESSAYS, Emma Goldman. Powerful, penetrating, prophetic essays on direct action, role of minorities, prison reform, puritan hypocrisy, violence, etc. 271pp. 5⅜ x 8½. 22484-8 Pa. $7.95

MYTHS OF THE HINDUS AND BUDDHISTS, Ananda K. Coomaraswamy and Sister Nivedita. Great stories of the epics; deeds of Krishna, Shiva, taken from puranas, Vedas, folk tales; etc. 32 illustrations. 400pp. 5⅜ x 8½. 21759-0 Pa. $12.95

THE TRAUMA OF BIRTH, Otto Rank. Rank's controversial thesis that anxiety neurosis is caused by profound psychological trauma which occurs at birth. 256pp. 5⅜ x 8½. 27974-X Pa. $7.95

A THEOLOGICO-POLITICAL TREATISE, Benedict Spinoza. Also contains unfinished Political Treatise. Great classic on religious liberty, theory of government on common consent. R. Elwes translation. Total of 421pp. 5⅜ x 8½. 20249-6 Pa. $9.95

CATALOG OF DOVER BOOKS

MY BONDAGE AND MY FREEDOM, Frederick Douglass. Born a slave, Douglass became outspoken force in antislavery movement. The best of Douglass' autobiographies. Graphic description of slave life. 464pp. 5⅜ x 8½. 22457-0 Pa. $8.95

FOLLOWING THE EQUATOR: A Journey Around the World, Mark Twain. Fascinating humorous account of 1897 voyage to Hawaii, Australia, India, New Zealand, etc. Ironic, bemused reports on peoples, customs, climate, flora and fauna, politics, much more. 197 illustrations. 720pp. 5⅜ x 8½. 26113-1 Pa. $15.95

THE PEOPLE CALLED SHAKERS, Edward D. Andrews. Definitive study of Shakers: origins, beliefs, practices, dances, social organization, furniture and crafts, etc. 33 illustrations. 351pp. 5⅜ x 8½. 21081-2 Pa. $8.95

THE MYTHS OF GREECE AND ROME, H. A. Guerber. A classic of mythology, generously illustrated, long prized for its simple, graphic, accurate retelling of the principal myths of Greece and Rome, and for its commentary on their origins and significance. With 64 illustrations by Michelangelo, Raphael, Titian, Rubens, Canova, Bernini and others. 480pp. 5⅜ x 8½. 27584-1 Pa. $9.95

PSYCHOLOGY OF MUSIC, Carl E. Seashore. Classic work discusses music as a medium from psychological viewpoint. Clear treatment of physical acoustics, auditory apparatus, sound perception, development of musical skills, nature of musical feeling, host of other topics. 88 figures. 408pp. 5⅜ x 8½. 21851-1 Pa. $11.95

THE PHILOSOPHY OF HISTORY, Georg W. Hegel. Great classic of Western thought develops concept that history is not chance but rational process, the evolution of freedom. 457pp. 5⅜ x 8½. 20112-0 Pa. $9.95

THE BOOK OF TEA, Kakuzo Okakura. Minor classic of the Orient: entertaining, charming explanation, interpretation of traditional Japanese culture in terms of tea ceremony. 94pp. 5⅜ x 8½. 20070-1 Pa. $3.95

LIFE IN ANCIENT EGYPT, Adolf Erman. Fullest, most thorough, detailed older account with much not in more recent books, domestic life, religion, magic, medicine, commerce, much more. Many illustrations reproduce tomb paintings, carvings, hieroglyphs, etc. 597pp. 5⅜ x 8½. 22632-8 Pa. $12.95

SUNDIALS, Their Theory and Construction, Albert Waugh. Far and away the best, most thorough coverage of ideas, mathematics concerned, types, construction, adjusting anywhere. Simple, nontechnical treatment allows even children to build several of these dials. Over 100 illustrations. 230pp. 5⅜ x 8½. 22947-5 Pa. $8.95

THEORETICAL HYDRODYNAMICS, L. M. Milne-Thomson. Classic exposition of the mathematical theory of fluid motion, applicable to both hydrodynamics and aerodynamics. Over 600 exercises. 768pp. 6⅛ x 9¼. 68970-0 Pa. $20.95

SONGS OF EXPERIENCE: Facsimile Reproduction with 26 Plates in Full Color, William Blake. 26 full-color plates from a rare 1826 edition. Includes "The Tyger," "London," "Holy Thursday," and other poems. Printed text of poems. 48pp. 5¼ x 7. 24636-1 Pa. $4.95

OLD-TIME VIGNETTES IN FULL COLOR, Carol Belanger Grafton (ed.). Over 390 charming, often sentimental illustrations, selected from archives of Victorian graphics—pretty women posing, children playing, food, flowers, kittens and puppies, smiling cherubs, birds and butterflies, much more. All copyright-free. 48pp. 9¼ x 12¼. 27269-9 Pa. $7.95

PERSPECTIVE FOR ARTISTS, Rex Vicat Cole. Depth, perspective of sky and sea, shadows, much more, not usually covered. 391 diagrams, 81 reproductions of drawings and paintings. 279pp. 5⅜ x 8½. 22487-2 Pa. $7.95

DRAWING THE LIVING FIGURE, Joseph Sheppard. Innovative approach to artistic anatomy focuses on specifics of surface anatomy, rather than muscles and bones. Over 170 drawings of live models in front, back and side views, and in widely varying poses. Accompanying diagrams. 177 illustrations. Introduction. Index. 144pp. 8⅜ x11¼. 26723-7 Pa. $8.95

GOTHIC AND OLD ENGLISH ALPHABETS: 100 Complete Fonts, Dan X. Solo. Add power, elegance to posters, signs, other graphics with 100 stunning copyright-free alphabets: Blackstone, Dolbey, Germania, 97 more—including many lower-case, numerals, punctuation marks. 104pp. 8⅛ x 11. 24695-7 Pa. $8.95

HOW TO DO BEADWORK, Mary White. Fundamental book on craft from simple projects to five-bead chains and woven works. 106 illustrations. 142pp. 5⅜ x 8. 20697-1 Pa. $5.95

THE BOOK OF WOOD CARVING, Charles Marshall Sayers. Finest book for beginners discusses fundamentals and offers 34 designs. "Absolutely first rate . . . well thought out and well executed."—E. J. Tangerman. 118pp. 7¾ x 10⅝. 23654-4 Pa. $7.95

ILLUSTRATED CATALOG OF CIVIL WAR MILITARY GOODS: Union Army Weapons, Insignia, Uniform Accessories, and Other Equipment, Schuyler, Hartley, and Graham. Rare, profusely illustrated 1846 catalog includes Union Army uniform and dress regulations, arms and ammunition, coats, insignia, flags, swords, rifles, etc. 226 illustrations. 160pp. 9 x 12. 24939-5 Pa. $10.95

WOMEN'S FASHIONS OF THE EARLY 1900s: An Unabridged Republication of "New York Fashions, 1909," National Cloak & Suit Co. Rare catalog of mail-order fashions documents women's and children's clothing styles shortly after the turn of the century. Captions offer full descriptions, prices. Invaluable resource for fashion, costume historians. Approximately 725 illustrations. 128pp. 8⅜ x 11¼. 27276-1 Pa. $11.95

THE 1912 AND 1915 GUSTAV STICKLEY FURNITURE CATALOGS, Gustav Stickley. With over 200 detailed illustrations and descriptions, these two catalogs are essential reading and reference materials and identification guides for Stickley furniture. Captions cite materials, dimensions and prices. 112pp. 6½ x 9¼. 26676-1 Pa. $9.95

EARLY AMERICAN LOCOMOTIVES, John H. White, Jr. Finest locomotive engravings from early 19th century: historical (1804–74), main-line (after 1870), special, foreign, etc. 147 plates. 142pp. 11⅜ x 8¼. 22772-3 Pa. $10.95

THE TALL SHIPS OF TODAY IN PHOTOGRAPHS, Frank O. Braynard. Lavishly illustrated tribute to nearly 100 majestic contemporary sailing vessels: Amerigo Vespucci, Clearwater, Constitution, Eagle, Mayflower, Sea Cloud, Victory, many more. Authoritative captions provide statistics, background on each ship. 190 black-and-white photographs and illustrations. Introduction. 128pp. 8⅞ x 11¾. 27163-3 Pa. $14.95

LITTLE BOOK OF EARLY AMERICAN CRAFTS AND TRADES, Peter Stockham (ed.). 1807 children's book explains crafts and trades: baker, hatter, cooper, potter, and many others. 23 copperplate illustrations. 140pp. $4\frac{5}{8}$ x 6.
23336-7 Pa. $4.95

VICTORIAN FASHIONS AND COSTUMES FROM HARPER'S BAZAR, 1867–1898, Stella Blum (ed.). Day costumes, evening wear, sports clothes, shoes, hats, other accessories in over 1,000 detailed engravings. 320pp. $9\frac{3}{8}$ x $12\frac{1}{4}$.
22990-4 Pa. $15.95

GUSTAV STICKLEY, THE CRAFTSMAN, Mary Ann Smith. Superb study surveys broad scope of Stickley's achievement, especially in architecture. Design philosophy, rise and fall of the Craftsman empire, descriptions and floor plans for many Craftsman houses, more. 86 black-and-white halftones. 31 line illustrations. Introduction 208pp. $6\frac{1}{2}$ x $9\frac{1}{4}$. 27210-9 Pa. $9.95

THE LONG ISLAND RAIL ROAD IN EARLY PHOTOGRAPHS, Ron Ziel. Over 220 rare photos, informative text document origin (1844) and development of rail service on Long Island. Vintage views of early trains, locomotives, stations, passengers, crews, much more. Captions. $8\frac{7}{8}$ x $11\frac{3}{4}$. 26301-0 Pa. $13.95

VOYAGE OF THE LIBERDADE, Joshua Slocum. Great 19th-century mariner's thrilling, first-hand account of the wreck of his ship off South America, the 35-foot boat he built from the wreckage, and its remarkable voyage home. 128pp. $5\frac{3}{8}$ x $8\frac{1}{2}$.
40022-0 Pa. $4.95

TEN BOOKS ON ARCHITECTURE, Vitruvius. The most important book ever written on architecture. Early Roman aesthetics, technology, classical orders, site selection, all other aspects. Morgan translation. 331pp. $5\frac{3}{8}$ x $8\frac{1}{2}$. 20645-9 Pa. $8.95

THE HUMAN FIGURE IN MOTION, Eadweard Muybridge. More than 4,500 stopped-action photos, in action series, showing undraped men, women, children jumping, lying down, throwing, sitting, wrestling, carrying, etc. 390pp. $7\frac{7}{8}$ x $10\frac{5}{8}$.
20204-6 Clothbd. $27.95

TREES OF THE EASTERN AND CENTRAL UNITED STATES AND CANADA, William M. Harlow. Best one-volume guide to 140 trees. Full descriptions, woodlore, range, etc. Over 600 illustrations. Handy size. 288pp. $4\frac{1}{2}$ x $6\frac{3}{8}$.
20395-6 Pa. $6.95

SONGS OF WESTERN BIRDS, Dr. Donald J. Borror. Complete song and call repertoire of 60 western species, including flycatchers, juncoes, cactus wrens, many more–includes fully illustrated booklet. Cassette and manual 99913-0 $8.95

GROWING AND USING HERBS AND SPICES, Milo Miloradovich. Versatile handbook provides all the information needed for cultivation and use of all the herbs and spices available in North America. 4 illustrations. Index. Glossary. 236pp. $5\frac{3}{8}$ x $8\frac{1}{2}$.
25058-X Pa. $7.95

BIG BOOK OF MAZES AND LABYRINTHS, Walter Shepherd. 50 mazes and labyrinths in all–classical, solid, ripple, and more–in one great volume. Perfect inexpensive puzzler for clever youngsters. Full solutions. 112pp. $8\frac{1}{8}$ x 11.
22951-3 Pa. $5.95

PIANO TUNING, J. Cree Fischer. Clearest, best book for beginner, amateur. Simple repairs, raising dropped notes, tuning by easy method of flattened fifths. No previous skills needed. 4 illustrations. 201pp. 5⅜ x 8½. 23267-0 Pa. $6.95

HINTS TO SINGERS, Lillian Nordica. Selecting the right teacher, developing confidence, overcoming stage fright, and many other important skills receive thoughtful discussion in this indispensible guide, written by a world-famous diva of four decades' experience. 96pp. 5³/₈ x 8¹/₂. 40094-8 Pa. $4.95

THE COMPLETE NONSENSE OF EDWARD LEAR, Edward Lear. All nonsense limericks, zany alphabets, Owl and Pussycat, songs, nonsense botany, etc., illustrated by Lear. Total of 320pp. 5⅜ x 8½. (USO) 20167-8 Pa. $7.95

VICTORIAN PARLOUR POETRY: An Annotated Anthology, Michael R. Turner. 117 gems by Longfellow, Tennyson, Browning, many lesser-known poets. "The Village Blacksmith," "Curfew Must Not Ring Tonight," "Only a Baby Small," dozens more, often difficult to find elsewhere. Index of poets, titles, first lines. xxiii + 325pp. 5⅜ x 8¼. 27044-0 Pa. $8.95

DUBLINERS, James Joyce. Fifteen stories offer vivid, tightly focused observations of the lives of Dublin's poorer classes. At least one, "The Dead," is considered a masterpiece. Reprinted complete and unabridged from standard edition. 160pp. 5³⁄₁₆ x 8¼. 26870-5 Pa. $1.00

GREAT WEIRD TALES: 14 Stories by Lovecraft, Blackwood, Machen and Others, S. T. Joshi (ed.). 14 spellbinding tales, including "The Sin Eater," by Fiona McLeod, "The Eye Above the Mantel," by Frank Belknap Long, as well as renowned works by R. H. Barlow, Lord Dunsany, Arthur Machen, W. C. Morrow and eight other masters of the genre. 256pp. 5⅜ x 8½. (USO) 40436-6 Pa. $8.95

THE BOOK OF THE SACRED MAGIC OF ABRAMELIN THE MAGE, translated by S. MacGregor Mathers. Medieval manuscript of ceremonial magic. Basic document in Aleister Crowley, Golden Dawn groups. 268pp. 5⅜ x 8½. 23211-5 Pa. $9.95

NEW RUSSIAN-ENGLISH AND ENGLISH-RUSSIAN DICTIONARY, M. A. O'Brien. This is a remarkably handy Russian dictionary, containing a surprising amount of information, including over 70,000 entries. 366pp. 4½ x 6⅛. 20208-9 Pa. $10.95

HISTORIC HOMES OF THE AMERICAN PRESIDENTS, Second, Revised Edition, Irvin Haas. A traveler's guide to American Presidential homes, most open to the public, depicting and describing homes occupied by every American President from George Washington to George Bush. With visiting hours, admission charges, travel routes. 175 photographs. Index. 160pp. 8¼ x 11. 26751-2 Pa. $11.95

NEW YORK IN THE FORTIES, Andreas Feininger. 162 brilliant photographs by the well-known photographer, formerly with *Life* magazine. Commuters, shoppers, Times Square at night, much else from city at its peak. Captions by John von Hartz. 181pp. 9¼ x 10¾. 23585-8 Pa. $13.95

INDIAN SIGN LANGUAGE, William Tomkins. Over 525 signs developed by Sioux and other tribes. Written instructions and diagrams. Also 290 pictographs. 111pp. 6⅛ x 9¼. 22029-X Pa. $3.95

ANATOMY: A Complete Guide for Artists, Joseph Sheppard. A master of figure drawing shows artists how to render human anatomy convincingly. Over 460 illustrations. 224pp. 8⅜ x 11¼. 27279-6 Pa. $11.95

MEDIEVAL CALLIGRAPHY: Its History and Technique, Marc Drogin. Spirited history, comprehensive instruction manual covers 13 styles (ca. 4th century thru 15th). Excellent photographs; directions for duplicating medieval techniques with modern tools. 224pp. 8⅜ x 11¼. 26142-5 Pa. $12.95

DRIED FLOWERS: How to Prepare Them, Sarah Whitlock and Martha Rankin. Complete instructions on how to use silica gel, meal and borax, perlite aggregate, sand and borax, glycerine and water to create attractive permanent flower arrangements. 12 illustrations. 32pp. 5⅜ x 8½. 21802-3 Pa. $1.00

EASY-TO-MAKE BIRD FEEDERS FOR WOODWORKERS, Scott D. Campbell. Detailed, simple-to-use guide for designing, constructing, caring for and using feeders. Text, illustrations for 12 classic and contemporary designs. 96pp. 5⅜ x 8½. 25847-5 Pa. $3.95

SCOTTISH WONDER TALES FROM MYTH AND LEGEND, Donald A. Mackenzie. 16 lively tales tell of giants rumbling down mountainsides, of a magic wand that turns stone pillars into warriors, of gods and goddesses, evil hags, powerful forces and more. 240pp. 5⅜ x 8½. 29677-6 Pa. $6.95

THE HISTORY OF UNDERCLOTHES, C. Willett Cunnington and Phyllis Cunnington. Fascinating, well-documented survey covering six centuries of English undergarments, enhanced with over 100 illustrations: 12th-century laced-up bodice, footed long drawers (1795), 19th-century bustles, 19th-century corsets for men, Victorian "bust improvers," much more. 272pp. 5⅝ x 8¼. 27124-2 Pa. $9.95

ARTS AND CRAFTS FURNITURE: The Complete Brooks Catalog of 1912, Brooks Manufacturing Co. Photos and detailed descriptions of more than 150 now very collectible furniture designs from the Arts and Crafts movement depict davenports, settees, buffets, desks, tables, chairs, bedsteads, dressers and more, all built of solid, quarter-sawed oak. Invaluable for students and enthusiasts of antiques, Americana and the decorative arts. 80pp. 6½ x 9¼. 27471-3 Pa. $8.95

WILBUR AND ORVILLE: A Biography of the Wright Brothers, Fred Howard. Definitive, crisply written study tells the full story of the brothers' lives and work. A vividly written biography, unparalleled in scope and color, that also captures the spirit of an extraordinary era. 560pp. 6⅛ x 9¼. 40297-5 Pa. $17.95

THE ARTS OF THE SAILOR: Knotting, Splicing and Ropework, Hervey Garrett Smith. Indispensable shipboard reference covers tools, basic knots and useful hitches; handsewing and canvas work, more. Over 100 illustrations. Delightful reading for sea lovers. 256pp. 5⅜ x 8½. 26440-8 Pa. $8.95

FRANK LLOYD WRIGHT'S FALLINGWATER: The House and Its History, Second, Revised Edition, Donald Hoffmann. A total revision–both in text and illustrations–of the standard document on Fallingwater, the boldest, most personal architectural statement of Wright's mature years, updated with valuable new material from the recently opened Frank Lloyd Wright Archives. "Fascinating"–*The New York Times*. 116 illustrations. 128pp. 9¼ x 10¾. 27430-6 Pa. $12.95

PHOTOGRAPHIC SKETCHBOOK OF THE CIVIL WAR, Alexander Gardner. 100 photos taken on field during the Civil War. Famous shots of Manassas Harper's Ferry, Lincoln, Richmond, slave pens, etc. 244pp. 10⅝ x 8¼. 22731-6 Pa. $10.95

FIVE ACRES AND INDEPENDENCE, Maurice G. Kains. Great back-to-the-land classic explains basics of self-sufficient farming. The one book to get. 95 illustrations. 397pp. 5⅜ x 8½. 20974-1 Pa. $7.95

SONGS OF EASTERN BIRDS, Dr. Donald J. Borror. Songs and calls of 60 species most common to eastern U.S.: warblers, woodpeckers, flycatchers, thrushes, larks, many more in high-quality recording. Cassette and manual 99912-2 $9.95

A MODERN HERBAL, Margaret Grieve. Much the fullest, most exact, most useful compilation of herbal material. Gigantic alphabetical encyclopedia, from aconite to zedoary, gives botanical information, medical properties, folklore, economic uses, much else. Indispensable to serious reader. 161 illustrations. 888pp. 6½ x 9¼. 2-vol. set. (USO) Vol. I: 22798-7 Pa. $9.95
Vol. II: 22799-5 Pa. $9.95

HIDDEN TREASURE MAZE BOOK, Dave Phillips. Solve 34 challenging mazes accompanied by heroic tales of adventure. Evil dragons, people-eating plants, blood-thirsty giants, many more dangerous adversaries lurk at every twist and turn. 34 mazes, stories, solutions. 48pp. 8¼ x 11. 24566-7 Pa. $2.95

LETTERS OF W. A. MOZART, Wolfgang A. Mozart. Remarkable letters show bawdy wit, humor, imagination, musical insights, contemporary musical world; includes some letters from Leopold Mozart. 276pp. 5⅜ x 8½. 22859-2 Pa. $7.95

BASIC PRINCIPLES OF CLASSICAL BALLET, Agrippina Vaganova. Great Russian theoretician, teacher explains methods for teaching classical ballet. 118 illustrations. 175pp. 5⅜ x 8½. 22036-2 Pa. $5.95

THE JUMPING FROG, Mark Twain. Revenge edition. The original story of The Celebrated Jumping Frog of Calaveras County, a hapless French translation, and Twain's hilarious "retranslation" from the French. 12 illustrations. 66pp. 5⅜ x 8½. 22686-7 Pa. $3.95

BEST REMEMBERED POEMS, Martin Gardner (ed.). The 126 poems in this superb collection of 19th- and 20th-century British and American verse range from Shelley's "To a Skylark" to the impassioned "Renascence" of Edna St. Vincent Millay and to Edward Lear's whimsical "The Owl and the Pussycat." 224pp. 5⅜ x 8½. 27165-X Pa. $5.95

COMPLETE SONNETS, William Shakespeare. Over 150 exquisite poems deal with love, friendship, the tyranny of time, beauty's evanescence, death and other themes in language of remarkable power, precision and beauty. Glossary of archaic terms. 80pp. 5³⁄₁₆ x 8¼. 26686-9 Pa. $1.00

BODIES IN A BOOKSHOP, R. T. Campbell. Challenging mystery of blackmail and murder with ingenious plot and superbly drawn characters. In the best tradition of British suspense fiction. 192pp. 5⅜ x 8½. 24720-1 Pa. $6.95

THE WIT AND HUMOR OF OSCAR WILDE, Alvin Redman (ed.). More than 1,000 ripostes, paradoxes, wisecracks: Work is the curse of the drinking classes; I can resist everything except temptation; etc. 258pp. 5⅜ x 8½. 20602-5 Pa. $6.95

SHAKESPEARE LEXICON AND QUOTATION DICTIONARY, Alexander Schmidt. Full definitions, locations, shades of meaning in every word in plays and poems. More than 50,000 exact quotations. 1,485pp. 6½ x 9¼. 2-vol. set.
Vol. 1: 22726-X Pa. $17.95
Vol. 2: 22727-8 Pa. $17.95

SELECTED POEMS, Emily Dickinson. Over 100 best-known, best-loved poems by one of America's foremost poets, reprinted from authoritative early editions. No comparable edition at this price. Index of first lines. 64pp. 5³⁄₁₆ x 8¼.
26466-1 Pa. $1.00

THE INSIDIOUS DR. FU-MANCHU, Sax Rohmer. The first of the popular mystery series introduces a pair of English detectives to their archnemesis, the diabolical Dr. Fu-Manchu. Flavorful atmosphere, fast-paced action, and colorful characters enliven this classic of the genre. 208pp. 5³⁄₁₆ x 8¼. 29898-1 Pa. $2.00

THE MALLEUS MALEFICARUM OF KRAMER AND SPRENGER, translated by Montague Summers. Full text of most important witchhunter's "bible," used by both Catholics and Protestants. 278pp. 6⅝ x 10. 22802-9 Pa. $12.95

SPANISH STORIES/CUENTOS ESPAÑOLES: A Dual-Language Book, Angel Flores (ed.). Unique format offers 13 great stories in Spanish by Cervantes, Borges, others. Faithful English translations on facing pages. 352pp. 5⅜ x 8½.
25399-6 Pa. $8.95

GARDEN CITY, LONG ISLAND, IN EARLY PHOTOGRAPHS, 1869–1919, Mildred H. Smith. Handsome treasury of 118 vintage pictures, accompanied by carefully researched captions, document the Garden City Hotel fire (1899), the Vanderbilt Cup Race (1908), the first airmail flight departing from the Nassau Boulevard Aerodrome (1911), and much more. 96pp. 8⅞ x 11¾. 40669-5 Pa. $12.95

OLD QUEENS, N.Y., IN EARLY PHOTOGRAPHS, Vincent F. Seyfried and William Asadorian. Over 160 rare photographs of Maspeth, Jamaica, Jackson Heights, and other areas. Vintage views of DeWitt Clinton mansion, 1939 World's Fair and more. Captions. 192pp. 8⅞ x 11. 26358-4 Pa. $12.95

CAPTURED BY THE INDIANS: 15 Firsthand Accounts, 1750-1870, Frederick Drimmer. Astounding true historical accounts of grisly torture, bloody conflicts, relentless pursuits, miraculous escapes and more, by people who lived to tell the tale. 384pp. 5⅜ x 8½. 24901-8 Pa. $8.95

THE WORLD'S GREAT SPEECHES (Fourth Enlarged Edition), Lewis Copeland, Lawrence W. Lamm, and Stephen J. McKenna. Nearly 300 speeches provide public speakers with a wealth of updated quotes and inspiration–from Pericles' funeral oration and William Jennings Bryan's "Cross of Gold Speech" to Malcolm X's powerful words on the Black Revolution and Earl of Spenser's tribute to his sister, Diana, Princess of Wales. 944pp. 5⅜ x 8⅜. 40903-1 Pa. $15.95

THE BOOK OF THE SWORD, Sir Richard F. Burton. Great Victorian scholar/adventurer's eloquent, erudite history of the "queen of weapons"–from prehistory to early Roman Empire. Evolution and development of early swords, variations (sabre, broadsword, cutlass, scimitar, etc.), much more. 336pp. 6⅛ x 9¼.
25434-8 Pa. $9.95

AUTOBIOGRAPHY: The Story of My Experiments with Truth, Mohandas K. Gandhi. Boyhood, legal studies, purification, the growth of the Satyagraha (nonviolent protest) movement. Critical, inspiring work of the man responsible for the freedom of India. 480pp. 5⅜ x 8½. (USO)   24593-4 Pa. $8.95

CELTIC MYTHS AND LEGENDS, T. W. Rolleston. Masterful retelling of Irish and Welsh stories and tales. Cuchulain, King Arthur, Deirdre, the Grail, many more. First paperback edition. 58 full-page illustrations. 512pp. 5⅜ x 8½.   26507-2 Pa. $9.95

THE PRINCIPLES OF PSYCHOLOGY, William James. Famous long course complete, unabridged. Stream of thought, time perception, memory, experimental methods; great work decades ahead of its time. 94 figures. 1,391pp. 5⅜ x 8½. 2-vol. set.
Vol. I: 20381-6 Pa. $13.95
Vol. II: 20382-4 Pa. $14.95

THE WORLD AS WILL AND REPRESENTATION, Arthur Schopenhauer. Definitive English translation of Schopenhauer's life work, correcting more than 1,000 errors, omissions in earlier translations. Translated by E. F. J. Payne. Total of 1,269pp. 5⅜ x 8½. 2-vol. set.
Vol. 1: 21761-2 Pa. $12.95
Vol. 2: 21762-0 Pa. $12.95

MAGIC AND MYSTERY IN TIBET, Madame Alexandra David-Neel. Experiences among lamas, magicians, sages, sorcerers, Bonpa wizards. A true psychic discovery. 32 illustrations. 321pp. 5⅜ x 8½. (USO)   22682-4 Pa. $9.95

THE EGYPTIAN BOOK OF THE DEAD, E. A. Wallis Budge. Complete reproduction of Ani's papyrus, finest ever found. Full hieroglyphic text, interlinear transliteration, word-for-word translation, smooth translation. 533pp. 6½ x 9¼.
21866-X Pa. $11.95

MATHEMATICS FOR THE NONMATHEMATICIAN, Morris Kline. Detailed, college-level treatment of mathematics in cultural and historical context, with numerous exercises. Recommended Reading Lists. Tables. Numerous figures. 641pp. 5⅜ x 8½.
24823-2 Pa. $11.95

PROBABILISTIC METHODS IN THE THEORY OF STRUCTURES, Isaac Elishakoff. Well-written introduction covers the elements of the theory of probability from two or more random variables, the reliability of such multivariable structures, the theory of random function, Monte Carlo methods of treating problems incapable of exact solution, and more. Examples. 502pp. 5³/₈ x 8¹/₂.   40691-1 Pa. $16.95

THE RIME OF THE ANCIENT MARINER, Gustave Doré, S. T. Coleridge. Doré's finest work; 34 plates capture moods, subtleties of poem. Flawless full-size reproductions printed on facing pages with authoritative text of poem. "Beautiful. Simply beautiful."—*Publisher's Weekly.* 77pp. 9¼ x 12.   22305-1 Pa. $7.95

NORTH AMERICAN INDIAN DESIGNS FOR ARTISTS AND CRAFTSPEOPLE, Eva Wilson. Over 360 authentic copyright-free designs adapted from Navajo blankets, Hopi pottery, Sioux buffalo hides, more. Geometrics, symbolic figures, plant and animal motifs, etc. 128pp. 8⅜ x 11. (EUK)   25341-4 Pa. $8.95

SCULPTURE: Principles and Practice, Louis Slobodkin. Step-by-step approach to clay, plaster, metals, stone; classical and modern. 253 drawings, photos. 255pp. 8⅛ x 11.
22960-2 Pa. $11.95

THE INFLUENCE OF SEA POWER UPON HISTORY, 1660–1783, A. T. Mahan. Influential classic of naval history and tactics still used as text in war colleges. First paperback edition. 4 maps. 24 battle plans. 640pp. 5⅜ x 8½. 25509-3 Pa. $14.95

THE STORY OF THE TITANIC AS TOLD BY ITS SURVIVORS, Jack Winocour (ed.). What it was really like. Panic, despair, shocking inefficiency, and a little heroism. More thrilling than any fictional account. 26 illustrations. 320pp. 5⅜ x 8½. 20610-6 Pa. $8.95

FAIRY AND FOLK TALES OF THE IRISH PEASANTRY, William Butler Yeats (ed.). Treasury of 64 tales from the twilight world of Celtic myth and legend: "The Soul Cages," "The Kildare Pooka," "King O'Toole and his Goose," many more. Introduction and Notes by W. B. Yeats. 352pp. 5⅜ x 8½. 26941-8 Pa. $8.95

BUDDHIST MAHAYANA TEXTS, E. B. Cowell and Others (eds.). Superb, accurate translations of basic documents in Mahayana Buddhism, highly important in history of religions. The Buddha-karita of Asvaghosha, Larger Sukhavativyuha, more. 448pp. 5⅜ x 8½. 25552-2 Pa. $12.95

ONE TWO THREE . . . INFINITY: Facts and Speculations of Science, George Gamow. Great physicist's fascinating, readable overview of contemporary science: number theory, relativity, fourth dimension, entropy, genes, atomic structure, much more. 128 illustrations. Index. 352pp. 5⅜ x 8½. 25664-2 Pa. $8.95

EXPERIMENTATION AND MEASUREMENT, W. J. Youden. Introductory manual explains laws of measurement in simple terms and offers tips for achieving accuracy and minimizing errors. Mathematics of measurement, use of instruments, experimenting with machines. 1994 edition. Foreword. Preface. Introduction. Epilogue. Selected Readings. Glossary. Index. Tables and figures. 128pp. 5³/₈ x 8¹/₂. 40451-X Pa. $6.95

DALÍ ON MODERN ART: The Cuckolds of Antiquated Modern Art, Salvador Dalí. Influential painter skewers modern art and its practitioners. Outrageous evaluations of Picasso, Cézanne, Turner, more. 15 renderings of paintings discussed. 44 calligraphic decorations by Dalí. 96pp. 5⅜ x 8½. (USO) 29220-7 Pa. $5.95

ANTIQUE PLAYING CARDS: A Pictorial History, Henry René D'Allemagne. Over 900 elaborate, decorative images from rare playing cards (14th–20th centuries): Bacchus, death, dancing dogs, hunting scenes, royal coats of arms, players cheating, much more. 96pp. 9¼ x 12¼. 29265-7 Pa. $12.95

MAKING FURNITURE MASTERPIECES: 30 Projects with Measured Drawings, Franklin H. Gottshall. Step-by-step instructions, illustrations for constructing handsome, useful pieces, among them a Sheraton desk, Chippendale chair, Spanish desk, Queen Anne table and a William and Mary dressing mirror. 224pp. 8⅛ x 11¼. 29338-6 Pa. $13.95

THE FOSSIL BOOK: A Record of Prehistoric Life, Patricia V. Rich et al. Profusely illustrated definitive guide covers everything from single-celled organisms and dinosaurs to birds and mammals and the interplay between climate and man. Over 1,500 illustrations. 760pp. 7½ x 10⅛. 29371-8 Pa. $29.95

Prices subject to change without notice.

Available at your book dealer or write for free catalog to Dept. GI, Dover Publications, Inc., 31 East 2nd St., Mineola, N.Y. 11501. Dover publishes more than 500 books each year on science, elementary and advanced mathematics, biology, music, art, literary history, social sciences and other areas.